The central purpose of this book is to illustrate the premiss that theoretical analysis of the kinetics of biological processes can give valuable information concerning the underlying mechanisms that are responsible for these processes.

Topics covered range from cooperativity in protein binding and enzyme action, through receptor–effector coupling, to theories of biochemical oscillations in yeast and slime mold, of liver regeneration, and of neurotransmitter release. Theories are always closely coupled to experiment.

The material of this book originally appeared as part of the volume *Mathematical models in molecular and cellular biology* (edited by Lee A. Segel). However, each chapter has been revised and updated.

CAMBRIDGE STUDIES IN MATHEMATICAL BIOLOGY: 12

Editors

C. CANNINGS
Department of Probability and Statistics, University of Sheffield, UK

F.C. HOPPENSTEADT
College of Natural Sciences, Michigan State University, East Lansing, USA

L.A. SEGEL
Weizmann Institute of Science, Rehovot, Israel

BIOLOGICAL KINETICS

CAMBRIDGE STUDIES
IN MATHEMATICAL BIOLOGY

Edited by
LEE A. SEGEL

Henry and Bertha Benson Professor of Mathematics
The Weizmann Institute of Science, Rehovot, Israel

Biological kinetics

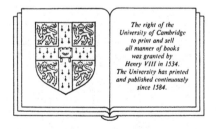

The right of the
University of Cambridge
to print and sell
all manner of books
was granted by
Henry VIII in 1534.
The University has printed
and published continuously
since 1584.

CAMBRIDGE UNIVERSITY PRESS
Cambridge
New York   Port Chester   Melbourne   Sydney

CAMBRIDGE UNIVERSITY PRESS
Cambridge, New York, Melbourne, Madrid, Cape Town, Singapore, São Paulo

Cambridge University Press
The Edinburgh Building, Cambridge CB2 2RU, UK

Published in the United States of America by Cambridge University Press, New York

www.cambridge.org
Information on this title: www.cambridge.org/9780521391849

First published 1991

A catalogue record for this publication is available from the British Library

Library of Congress Cataloguing in Publication data

Biological kinetics / edited by Lee A. Segel.
   p.   cm. — (Cambridge studies in mathematical biology : 12)
Includes index.
ISBN 0 521 39184 9
1. Molecular biology—Mathematical models.   2. Biophysics.
3. Molecular dynamics.   I. Segel, Lee A.   II. Series.
QH506.B543   1991
574.8′8′0151—dc20   90-2465   CIP

ISBN-13 978-0-521-39184-9 hardback
ISBN-10 0-521-39184-9 hardback

Transferred to digital printing 2006

CONTENTS

CONTRIBUTORS

Bard, Jonathan, B.L.   Western General Hospital, MRC
Human Genetics Unit, Crewe Road,
Edinburgh EH4 2XU, Scotland

Ghozlan, Aline   Pitie-Salpetriere Hospital, Department
of Psychology, Paris, France

Goldbeter, Albert   Université Libre de Bruxelles, Service de
Chimie Physique II, 1050 Bruxelles,
Belgium

Levitzki, Alexander   The Hebrew University of Jerusalem,
Department of Biological Chemistry,
Institute of Life Sciences, Jerusalem,
Israel

Parnas, Hanna   The Hebrew University of Jerusalem,
Department of Neurobiology,
Jerusalem, Israel

Perelson, Alan S.   Los Alamos National Laboratory, T-10
Division, Mail Stop K710, Los Alamos,
NM 87545, USA, and Santa Fe Institute,
1120 Canyon Road, Santa Fe,
NM 87501, USA

Rubinow, Sol, I.   Deceased

Segel, Lee A.   The Weizmann Institute of Science,
Department of Applied Mathematics
and Computer Science, Rehovot, Israel

Tolkovsky, Aviva M.   Oxford University, Department of
Neurobiology, Oxford, UK

Yagil, Gad   The Weizmann Institute of Science,
Department of Cell Biology, Rehovot,
Israel

PREFACE

The central purpose of this book is to illustrate the premiss that examination of the kinetics (time course) of biological processes can give valuable information concerning the underlying mechanisms that are responsible for these processes. To extract this information it is usually necessary to construct a mathematical model that embodies hypothesized mechanisms. Solution of the resulting equations shows whether the hypotheses are consistent with the data.

Considerable material concerns steady-state solutions. These can be regarded as the limiting behavior, in many instances, of the kinetic equations.

On the molecular level, the discourse ranges from fairly classical analyses of cooperativity in protein binding and enzyme action, through studies of enzyme induction and receptor–effector coupling, to theories for biochemical oscillations in yeast and slime mold. Models for the triggering of secretion in slime mold and in nerve cells, and for liver regeneration, are at the intersection of molecular biology, cellular biology and physiology. In addition, an introduction to the explosively growing theoretical topic of chaos concludes with references that chronicle tentative attempts to apply chaos theory in physiology (cardiac dynamics and immunology).

The material of this book originally appeared as part of *Mathematical models in molecular and cellular biology* (Lee A. Segel, ed., Cambridge: Cambridge University Press, 1980), which is now out of print. Each contribution has been revised and updated. (Unfortunately, Sol Rubinow has passed away. His contribution appears with permission of his widow, Shirley Rubinow, and has been updated by Lee A. Segel.)

The mathematical requisite for most of the material is a good command of basic calculus. A brief summary of the required mathematical ideas can be found in the Appendices of 'MDP', *Modelling dynamic phenomena in molecular and cellular biology* by Lee A. Segel (Cambridge: Cambridge University Press, 1984). Indeed, one of the uses to which the present book might be put is as a supplement to MDP, or to other texts in theoretical biology. There is substantial overlap with MDP in only one topic, cAMP secretion in slime mold. However, the coverage of this topic in the present volume – although perhaps less detailed mathematically – is more comprehensive and up-to-date.

It is hoped that this volume will be of interest to students and researchers alike, in both biology and applied mathematics. Readers should find a number of interesting case studies that show how mathematical modeling can illuminate important areas of modern biology.

1

Fundamental concepts in biochemical reaction theory

Law of mass action

Consider a reaction in which a chemical A of concentration A combines reversibly with a chemical B of concentration B to yield **complex**, C, of concentration C. This reaction is symbolized by

$$A + B \underset{k_{-1}}{\overset{k_{+1}}{\rightleftharpoons}} C. \tag{1}$$

The **forward and backward rate constants** k_{+1} and k_{-1} are the proportionality factors in the **law of mass action** that is assumed to describe the process of the reaction. According to this law, the rate at which the species A reacts to form C is proportional to the mass of A, or equivalently, to the number of molecules of A available for reaction. In mathematical terms, the law takes the form of the following differential equations for the concentrations A, B, and C, at time t;

$$dA/dt = -k_{+1}AB + k_{-1}C, \qquad dB/dt = -k_{+1}AB + k_{-1}C, \quad (2a, b)$$

$$dC/dt = k_{+1}AB - k_{-1}C. \tag{2c}$$

In (2a), A is supposed to decrease at a rate jointly proportional to the concentrations of A and B. The idea behind this is again the law of mass action: doubling the concentrations of either A or B will double the rate of collision between these two molecules and hence will double the rate of 'successful' collisions that lead to the formation of C. Such an assumption is plausible as long as the concentrations are not too large. The break-up of an individual C molecule into its constituents is held to occur with a constant probability per unit time.

The phenomenological law of mass action can, in principle, be derived from statistical mechanics, or on a deeper level from quantum mechanics, but this law can be regarded as being well established because of experimental information on a wide variety of theories in the biological, chemical and physical sciences that assume it.

Enzyme–substrate complex system

Enzymes are large molecules that speed up the conversion of a chemical to an altered form. According to the theory of enzymatic reactions of Michaelis & Menten (1913), the enzyme accomplishes this in two steps. First the enzyme (concentration E) reacts reversibly with the chemical, called a **substrate** in this context, to form a **complex** (concentration C). Secondly, the complex breaks apart into an altered substrate or **product** and the original enzyme. This last reaction is often assumed to be irreversible, in which case one writes

$$E + S \underset{k_{-1}}{\overset{k_{+1}}{\rightleftharpoons}} C \xrightarrow{k_{+2}} E + P.$$

The law of mass action for the concentrations $E(t)$, $S(t)$, $C(t)$, and $S(t)$ takes the form

$$dE/dt = -k_{+1}ES + k_{-1}C + k_{+2}C, \tag{3a}$$

$$dS/dt = -k_{+1}ES + k_{-1}C, \tag{3b}$$

$$dC/dt = k_{+1}ES - k_{-1}C - k_{+2}C, \tag{3c}$$

$$dP/dt = k_{+2}C. \tag{3d}$$

The above **system of differential equations** representing the enzymatic conversion of substrates to product was first put forward by Briggs & Haldane (1925). The equation must be supplemented by **initial conditions** that describe the system at some reference time. This time is conveniently designated $t = 0$. The standard initial conditions, which conform to the usual investigation of enzymatically controlled reactions, prescribe starting concentrations of enzyme and substrate, and assume that complex and product have had no opportunity to form:

$$E(0) = E_0, \qquad S(0) = S_0, \qquad C(0) = 0, \qquad P(0) = 0. \tag{4}$$

Addition of (3a) and (3c) yields

$$d(E + C)/dt = 0. \tag{5}$$

Consequently $E + C$ must be a constant, reflecting the fact that at any time t all enzyme molecules are either in their original form or bound in a complex. Using the initial conditions, the constant can be determined, so that we can write the **conservation equation**

$$E(t) + C(t) = E_0. \tag{6a}$$

This equation may be used to eliminate E from (3b) and (3c), leaving two equations for the two unknown functions $S(t)$ and $C(t)$:

$$\mathrm{d}S/\mathrm{d}t = k_{+1}E_0S + C(k_{+1}S + k_{-1}), \tag{6b}$$

$$\mathrm{d}C/\mathrm{d}t = k_{+1}E_0S - k_{+1}C(S + K_m), \tag{6c}$$

$$K_m \equiv (k_{-1} + k_{+2})/k_{+1}. \tag{6d}$$

Pseudo-steady state: Michaelis–Menten equation

In laboratory experiments, it is typically the case that, at the start, many substrate molecules are present for each enzyme molecule. Under these circumstances one expects that after an initial short transient period there will be a balance between the formation of complex by the union of enzyme and substrate and the breaking apart of complex (either to enzyme and substrate, or to enzyme and product). Because there are so many substrate molecules, this balance will be achieved before there is perceptible transformation of substrate into product. One anticipates, therefore, that calculation of product formation can be carried out under the assumption that $\mathrm{d}C/\mathrm{d}t = 0$, or, from (3c),

$$k_{+1}ES = (k_{-1} + k_{+2})C. \tag{7}$$

This equation is said to result from a **quasi-** or **pseudo-steady state hypothesis**. If any quantity no longer changes with time it is said to be in a **steady state**. We add 'pseudo' or 'quasi' to the description of (7) as a steady state, since although C is fully adjusted to the instantaneous values of E and S, those values are changing slowly with time.

Upon substitution of (6a) and (7) into (3b), we obtain the following equation for S:

$$\mathrm{d}S/\mathrm{d}t = -k_2E_0S/(K_m + S). \tag{8}$$

The solution of (8) (by the method of separation of variables) subject to the initial condition $S(0) = S_0$, is

$$S + \frac{k_{-1} + k_{+2}}{k_{+1}} \ln \frac{S}{S_0} = S_0 - k_{+2}E_0t.$$

Of particular interest is the velocity of reaction $V(t)$ defined as the rate of appearance of product. In view of the steady state hypothesis, we have from (6a), (7) and (3d) that

$$V(t) = \mathrm{d}P/\mathrm{d}t = k_{+2}C = |\mathrm{d}S/\mathrm{d}t|. \tag{9}$$

Biochemists are usually interested in $V(t)$ at the beginning of the reaction. From (8) we can write for this **initial velocity** $V_0 \equiv V(0)$,

$$V_0 = \frac{VS_0}{K_m + S_0},$$ \hfill (10a)

where

$$V \equiv k_{+2}E_0.$$ \hfill (10b)

Equation (10a) is called the **Michaelis–Menten** equation. Its graph starts from the origin, for the absence of substrate implies the absence of reaction, and approaches the asymptote $V_0 = V$ as S_0 becomes larger and larger (see Figure 1.1). Thus, when S_0 is large compared to K_m and $V_0 \approx V$, there is an abundance of substrate and the 'chemical factory' is working as fast as possible. In such cases the system is said to be **saturated**. Because the constant V is the maximum velocity that the reaction can attain, the term 'V-max' is used to describe it. (The term **Langmuir isotherm** is also associated with (10a), which is said to have the form of a **rectangular hyperbola**.)

The biochemical determination of the **Michaelis constant** K_m follows from the observation that when $S_0 = K_m$ then $V_0 = \frac{1}{2}V$. Thus K_m gives the concentration at which the reaction attains its half-maximal value. If this concentration is relatively low, then the reaction is said to be highly **specific**. A relatively low K_m means a relatively large k_{+1} and this in turn means that an enzyme–substrate collision is relatively likely to result in the formation of

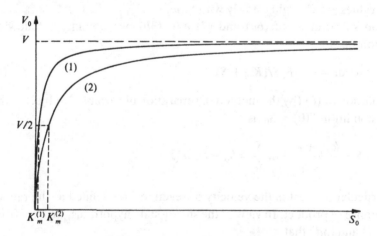

Figure 1.1. Graphs of the Michaelis–Menten equation (10a) in two situations with the same maximum velocity V. The reaction represented by curve (1) is more specific than that of curve (2) because the Michaelis constant for it is smaller:

$$K_m^{(1)} < K_m^{(2)}.$$

product, i.e. that the enzyme is specifically adapted to act on the particular substrate.

Biochemists frequently rearrange the Michaelis–Menten equation (10a) into the **Lineweaver–Burk** or **double-reciprocal** form

$$\frac{1}{V_0} = \frac{1}{V} + \left(\frac{K_m}{V}\right)\frac{1}{S_0}. \tag{11}$$

The graph $1/V_0$ versus $1/S_0$ is thus a straight line, which simplifies the problem of fitting the theory to data. Then $1/V$ and $-1/K_m$ can be found at once as the intersection of this line with the vertical and horizontal axes, respectively (Figure 1.2).

Note from (10b) that V depends on the product of the initial enzyme concentration E_0 and the product formation rate constant k_{+2}. This reflects the fact that at high substrate concentrations the speed of reaction depends only on how many reaction units there are, and on how fast they can transform complex into product. Under such circumstances one says that the enzyme is the **rate-limiting chemical** and the complex–product conversion is the **rate-limiting step** in the conversion of substrate to product.

The back reaction for the conversion of complex to product can also be included in the theory. Further, Haldane (1930) has indicated that the reaction of product and substrate should be viewed symmetrically so that the complete set of reactions presumed to take place is represented as

$$S + E \underset{k_{-1}}{\overset{k_{+1}}{\rightleftharpoons}} C \underset{k'_{-2}}{\overset{k'_{+2}}{\rightleftharpoons}} C' \underset{k_{-2}}{\overset{k_{+2}}{\rightleftharpoons}} P + E, \tag{12}$$

where C is an S–E complex, and C' is a P–E complex. The analysis of this reaction scheme does not alter the form of the Michaelis–Menten equation (10a), although the meanings of V and K_m in terms of fundamental rate constants are more complicated than indicated by (10b) and (6d).

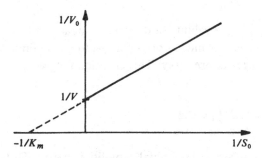

Figure 1.2. The Lineweaver–Burk plot, from which V and K_m can be readily determined. (The dashed part of the line corresponds to 'unphysical' negative substrate concentrations.)

The assumption of a pseudo-steady state can simplify a wide variety of kinetic problems. The most elementary application of this assumption yields the Michaelis–Menten equation (10a) that is a keystone of theoretical biochemistry. For both of these reasons it is worth carefully working out the conditions under which the pseudo-steady state assumption (7) is expected to be valid.

The key concept here is that of 'time scale', the order of magnitude of time that characterizes the duration of a process or subprocess. For example, what is the time scale of the fast transient process during which the complex concentration changes from its initial value of zero to a pseudo-state condition? Does it take microseconds, milliseconds or seconds? To *estimate* the duration of this period we can make the approximation $S = S_0$ in (6c). This transforms (6c) into a linear equation, with the solution

$$C(t) = \bar{C}[1 - \exp(-\mu t)], \qquad \mu \equiv k_{+1}(S_0 + K_m),$$
$$\bar{C} \equiv E_0 S_0/(K_m + S_0). \tag{13a, b, c}$$

Thus the *complex (fast) time scale* is given by $t_C = \mu^{-1}$:

$$t_C = [k_{+1}(S_0 + K_m)]^{-1}. \tag{14}$$

Now let us estimate the *substrate (slow) time scale* t_S, namely how long it takes for a significant change to occur in the substrate concentration. We employ the characterization (Segel 1984, p. 56).

$$t_S \approx \frac{\text{total change in } S \text{ after fast transient}}{\max |dS/dt| \text{ after fast transient}}. \tag{15}$$

The numerator of (15) is approximately S_0. Assuming the validity of the steady state assumption, we observe that the denominator is given by (8) with $S = S_0$. Thus $t_S \approx S_0/[k_{+2}E_0 S_0/(K_m + S_0)]$, i.e.

$$t_S = (K_m + S_0)/k_{+2}E_0. \tag{16}$$

One necessary criterion for the validity of the pseudo-steady state assumption is that the 'fast transient' is indeed brief compared to the time during which the substrate changes appreciably. This criterion is $t_C \ll t_S$ or, from (14) and (16)

$$\frac{E_0}{K_m + S_0} \ll \left(1 + \frac{k_{-1}}{k_{+2}}\right)\left(1 + \frac{S_0}{K_m}\right). \tag{17}$$

A second criterion concerns the 'initial' condition $S(0) = S_0$ that is imposed on (8). For this condition to be approximately valid there must be only a negligible decrease in substrate concentration during the duration t_C

of the brief transient. This decrease, which we denote by ΔS, is certainly less than the product of the time duration t_C and the initial (maximal) rate of substrate consumption. 'Initial' in the previous sentence refers to the very beginning of the experiment, so that the desired rate is obtained by setting $t = 0$ in (6*b*). This yields

$$\left| \frac{\Delta S}{S_0} \right| = \frac{1}{S_0} \left| \frac{dS}{dt} \right|_{max} \cdot t_C = \frac{E_0}{K_m + S_0}. \tag{18}$$

The requirement that $|\Delta S/S_0|$ be small compared to unity is thus expressed by

$$\varepsilon \ll 1, \qquad \text{where } \varepsilon \equiv \frac{E_0}{K_m + S_0}. \tag{19}$$

If (19) holds then (17) holds. Thus $\varepsilon \ll 1$ is a simple criterion for the validity of the pseudo-steady state assumption.

For considerable further discussion along the above lines see Segel (1988) and Segel & Slemrod (1989).

References

Briggs, G. E. & Haldane, J. B. S. (1925). A note on the kinetics of enzyme action. *Biochem. J.* **19**, 338–9.

Haldane, J. B. S. (1930). *Enzymes*, 2nd edn, London, Longmans, Green (reprinted by MIT Press, Cambridge, Mass. 1965), Chapter 5.

Michaelis, L. & Menten, M. L. (1913). Die kinetik der Invertinwirkung. *Biochem. Z.* **49**, 333–69.

Segel, L. A. (1984). *Modeling dynamic phenomena in molecular and cellular biology*, Cambridge, Cambridge University Press.

Segel, L. A. (1988). On the validity of the steady state assumption of enzyme kinetics. *Bull. Math. Biol.* **50**, 579–93.

Segel, L. A. & Slemrod, M. (1989). The quasi-steady state assumption: a case study in perturbation. *SIAM Rev.* **31**, 446–77.

2

Equilibrium binding of macromolecules with ligands

Theory of equilibrium dialysis

A basic and important method for studying the reaction of a protein P with a small molecule or ion C is **equilibrium dialysis**. In it, a known amount of the macromolecular protein is placed in solution inside a membrane bag that is suspended in a solution of the small molecule with which it is capable of reacting (the **ligand**). The membrane is permeable to the ligand, but impermeable to the macromolecule. Available membranes possessing such a permeability property require the molecular weight of the macromolecule to be greater than 10 000. The solution is allowed to stand for a sufficiently long time, of the order of one or two days, for the ligand to permeate the membrane and react with the protein so that equilibrium is established. At equilibrium, there exists both bound and unbound components of the ligand. Furthermore, the unbound concentrations on both sides of the membrane are equal to each other (if the ligand is an ion, electrical effects must be neutralized for the equality to hold true). By measuring the equilibrium concentrations outside the bag at the beginning and end of the equilibrium dialysis experiment, the bound ligand concentration is readily determined as the difference of those two quantities.

We present here the theory of this simple experimental procedure. We shall asume that the protein possesses n binding sites for the ligand, where n is an integer greater than unity in usual cases of interest. Let us denote by P_j the complex of a protein molecule with j ligand molecules attached, where $j = 0, 1, 2, \ldots, n$ (P_0 represents the bare protein). Then the reactions leading to equilibrium are as follows,

$$C + P_{j-1} \rightleftarrows P_j, \qquad j = 1, 2, \ldots, n. \tag{1}$$

Note that we have tacitly assumed that all complexes consisting of exactly j ligands attached are the same, regardless of the set of j attachment sites. We denote by italic lower case letters the concentrations of quantities represented by capital letters. Let us consider mathematically the first reaction above for $j = 1$. Employing the law of mass action, we express the time-dependent behavior of the concentration of one of these reactants, say p_0, as

$$dp_0/dt = -k_{+1}p_0c + k_{-1}p_1. \tag{2}$$

At equilibrium, p_0, c, and p_1 are no longer time dependent and attain constant values which are interrelated, according to (2), by the relation

$$0 = -k_{+1}p_0c + k_{-1}p_1. \tag{3}$$

It is customary to define the **association constant** K_a as

$$K_a \equiv k_{+1}/k_{-1}, \tag{4}$$

although for some purposes, it is found more convenient to utilize its inverse, the **dissociation constant** $K_d = 1/K_a = k_{-1}/k_{+1}$. Either one of these may be referred to as the **equilibrium constant**, although a certain ambiguity is thereby introduced by this usage if no further clarification is made. According to equations (3) and (4), at equilibrium

$$K_a = p_1/p_0c. \tag{5}$$

To describe now the full set of reactions (1) at equilibrium, we introduce, for uniformity of notation, the n association constants $K_j, j = 1, 2, \ldots, n$, by definitions analogous to (4). Generalizing (5)

$$K_1 = \frac{p_1}{cp_0}, K_2 = \frac{p_2}{cp_1}, \ldots,$$

$$K_j = \frac{p_j}{cp_{j-1}}, \ldots, K_n = \frac{p_n}{cp_{n-1}}. \tag{6}$$

The quantities $p_0, p_1, \ldots, p_n$ are not usually experimentally determinable, but it is possible to find the average number of molecules of C associated with each macromolecule. This number is denoted by r and is defined as

$$r \equiv \frac{\text{total number of molecules of C combined with P}}{\text{total number of molecules of P}}. \tag{7}$$

We shall call r the **mean association function**. The numerator and denominator above are experimentally measurable quantities, as already indicated. Because there are j ligand molecules attached to each molecule P_j, r is expressible as

$$r = \frac{p_1 + 2p_2 + 3p_3 + \ldots np_n}{p_0 + p_1 + p_2 + \ldots + p_n}, \tag{8}$$

or, using (6),

$$r = \frac{K_1 c + 2K_1 K_2 c^2 + 3K_1 K_2 K_3 c^3 + \ldots + nK_1 K_2 \ldots K_n c^n}{1 + K_1 c + K_1 K_2 c^2 + \ldots + K_1 K_2 \ldots K_n c^n}. \tag{9}$$

The above result is known as **Adair's equation** (Adair, 1925). A related quantity frequently utilized is the **saturation function** Y defined as the mean fraction of sites per protein molecule that are occupied, or

$$Y = r/n. \tag{10}$$

Identical independent sites

A significant simplification occurs when the binding sites in the protein are identical. Further, assume that binding at a given site is independent of the state of binding of all other sites. That is to say, let k_+ be the forward rate constant for attachment of a ligand molecule at a particular binding site, and let k_- be the associated backward rate constant. In terms of these rate constants, (3) assumes the form

$$0 = -nk_+ p_0 c + k_- p_1. \tag{11}$$

The factor n appears because there are n possible ways to form the state P_1 from the state P_0 (n available sites of attachment of the ligand molecule). Contrariwise, there is only one way for the ligand to be removed from the state P_1 to form the state P_0 (the ligand molecule is removed at its site of attachment). By a similar argument, we see that equilibrium between the states P_1 and P_2 is described as

$$0 = -(n - 1)k_+ p_1 c + 2k_- p_2, \tag{12}$$

i.e. there are $n - 1$ empty sites available for ligand binding in the state P_1, and two ways to remove a ligand molecule from the state P_2. Hence, with the **intrinsic association constant** K defined as

$$K \equiv k_+/k_-, \tag{13}$$

we infer that $K_1 = nK$, $K_2 = (n - 1)K/2$, and, in general (Bjerrum, 1941),

$$K_j = (n - j + 1)K/j, \qquad j = 1, 2, \ldots, n. \tag{14}$$

Thus, the assumption that the binding sites of the protein are identical and independent is equivalent to the assertion that the intrinsic binding constant at one site is the same as at any other site, and moreover is unaffected by the state of binding of the other sites. By (14), Adair's equation assumes the particular simple Michaelis–Menten form (see, for example, the derivation in Rubinow, 1975, Section 2.4)

$$r = \frac{nKc}{1 + Kc}.$$ (15)

A protein containing n binding sites which obeys equation (15) is said to be **noncooperative** or to display **zero cooperativity**. By contrast, if the protein obeys (9) for $n > 1$ and the K_j are not related by (14), it is said to be **cooperative** or to display **cooperativity**. (See also Chapter 3.)

Rather than repeat the cited derivation, we shall present here a simpler alternative derivation of (15) for the case of identical, independent binding sites. We focus on the individual binding sites and consider them as independent particles, as it were, because the fact that they are grouped in bundles of n on proteins has no bearing on their reaction properties. Let f be the equilibrium concentration of free binding sites, and b the equilibrium concentration of bound sites, which is the same as the equilibrium concentration of bound ligand. Then the intrinsic association constant K of a binding site satisfies

$$K = b/cf.$$ (16)

The total number of sites, whether free or bound, is expressed as

$$np = f + b.$$ (17)

According to the definition (7) of r, it is given as

$$r = b/p.$$ (18)

By dividing (17) by b, introducing (18), and eliminating f/b by means of (16), we see that (15) follows directly.

Hapten–antibody interactions and heterogeneity

The immunological system of higher organisms reacts to the stimulus of 'foreign' substances, called **antigens**, by producing certain large molecules, called **antibodies**. These combine with the antigens and thereby neutralize their harmful effect. The reaction of antibodies with antigens is chemical in nature. Antibodies are macromolecular proteins of blood serum of the order of 10^5 or 10^6 daltons in molecular weight, called **immunoglobulins**. Antigens normally have a molecular weight of 10^4 or larger. However, the formation of antibodies is a reaction to only a small part or chemical subunit of the antigen, called an **antigenic determinant**. Many low molecular weight chemical compounds have the ability to react with antibodies *in vitro*, and are called **haptens** in the vocabulary of immunology. However, they lack the ability to elicit the production of antibody *in vivo*. Such an ability can be conferred on them by attaching them to protein carriers, thereby producing hapten–carrier complexes which are considerably larger than the hapten molecules.

Antibody molecules contain many binding sites for haptens. Moreover, it is well known that antibody is **heterogeneous** in its affinity for antigen. This means that, given an antigen containing a single antigenic determinant, or a hapten that is reacting with antibody, there will be a range of association constants characterizing the hapten–antibody complex. Here we shall concern ourselves with a particular type of heterogeneity, namely that resulting from the macromolecule containing more than one set of binding sites for a given ligand. Such heterogeneity appears to be unique to the immune response, and is not seen for example in the behavior of enzymes reacting with substrate molecules.

Let us assume, for definiteness, that the macromolecule contains two sets of binding sites, each set consisting of identical, independent binding sites. Let the first set consist of n_1 sites having an intrinsic association constant K_1, with the subscript 2 designating the second set of sites. In analogy with the derivation of equation (15) we can write that, at equilibrium,

$$b_1/p = n_1 K_1 c/(1 + K_1 c), \qquad b_2/p = n_2 K_2 c/(1 + K_2 c). \tag{19}$$

The ratio of the number of bound sites to the number of proteins, i.e. the mean association function is given by r:

$$r = (b_1 + b_2)/p. \tag{20}$$

Hence, from (19),

$$r = \frac{n_1 K_1 c}{1 + K_1 c} + \frac{n_2 K_2 c}{1 + K_2 c}. \tag{21}$$

More generally, when there are m subpopulations of identical, independent binding sites, it follows by the same argument (Tanford, 1961) that the average number of occupied sites per protein molecule is

$$r = \sum_{j=1}^{m} \frac{n_j K_j c}{1 + K_j c}. \tag{22}$$

The Scatchard plot

For the case of a macromolecule containing a single set of n identical, independent binding sites, we recognize that the functional relationship between r and c given by (15) is the same as that between the initial velocity v of the reaction between an enzyme and a substrate of concentration s, namely (as in Eq. (1.10a))

$$v = Vs/(K_m + s),\tag{23}$$

where V and K_m are constants. The usual quantitative problem of characterizing such a reaction is to determine the constants V and K_m from the experimentally measured function $v(s)$. Similarly, we wish to determine n and K of (15), from the knowledge of $r(c)$. Analogous to the Lineweaver–Burk plot of enzyme kinetics (Lineweaver & Burk, 1934), we can plot $1/r$ as a function of $1/c$ (Klotz, 1946). Then if r satisfies (15), the relationship is linear, and the parameters n and K are readily found from the slope and intercept of the resulting straight line.

An alternative suggestion made by Scatchard (1949) is based on the rearrangement of (15) obtained by clearing the denominator and dividing by c so that (15) becomes

$$r/c = K(n - r).\tag{24}$$

Hence, a plot of the quantity r/c versus r is a straight line, as shown in Figure 2.1. Such a plot is called a **Scatchard plot**. For simplicity of notation, let us designate $s \equiv r/c$, a function of c, as the **Scatchard function**. We shall refer to the curve appearing in the Scatchard plot as the **Scatchard curve**. We note

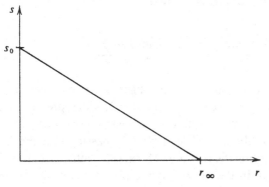

Figure 2.1. The Scatchard plot, based on (24) with $s_0 = nK$ and $r_\infty = n$.

parenthetically that the relationship analogous to (24) for representing the initial velocity of an enzymatic reaction reads

$$v/s = (1/K_m)(V - v), \tag{25}$$

and the corresponding plot of v/s versus v, called an **Eadie plot** (Eadie, 1942), was originally suggested by Woolf (1932).

We ask, with respect to Figure 2.1, how is the straight line traversed as c increases from zero to infinity? We see from (15) that r increases monotonically as c increases, so that the straight line evolves from upper left to lower right in Figure 2.1. In fact, from (15)

$$r \to 0 \quad \text{as } c \to 0,$$
$$r \to n \quad \text{as } c \to \infty. \tag{26}$$

At the same time, we see from (15) after division by c that

$$s \to nK \quad \text{as } c \to 0,$$
$$s \to 0 \quad \text{as } c \to \infty. \tag{27}$$

By means of (26) and (27), the values of the intercepts $r_\infty = n$ and $s_0 = nk$ of the Scatchard curve shown in Figure 2.1 are deduced.

The fact that the entire variation of r with c is represented in the Scatchard plot by a curve of finite length implies that experiments conducted over a large range of variation of c receive equal representation in the curve, so to speak. This property, lacking in the curve $r(c)$ which tends to emphasize small values of c, or in a Lineweaver–Burk plot which tends to emphasize large values of c, is an attractive feature of the Scatchard plot that has helped enhance its popularity in recent times.

What is the form of the Scatchard curve when r satisfies (22)? The analogs of the limits (27) and (26) are as follows:

$$s_0 \equiv \lim_{c \to 0} s = \sum_{i=1}^{m} n_i K_i, \qquad r_\infty \equiv \lim_{c \to \infty} r = \sum_{i=1}^{m} n_i. \tag{28}$$

However, the Scatchard curve is no longer a straight line. Figure 2.2 illustrates the Scatchard curve for the case $m = 2$, with representative values of the parameters for the binding of a ligand (1-iodonaphthalene- 4-sulfonate) to IgM antibodies. (We remark in passing that the Scatchard curve, when representing the functional form (22) with $n \geq 2$ always has the concave shape shown in the figure. By the phenomenological criterion of cooperativity (see Chapter 3), such curves indicate negative cooperative

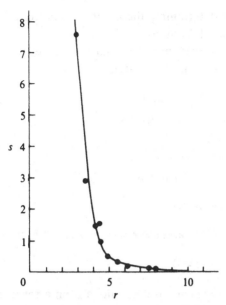

Figure 2.2. Scatchard plot representing the binding of a ligand (1-iodonaphthalene-4-sulfonate) by IgM antibodies. Solid circles are data points of Onoue *et al.* (1968). The curve is theoretical, and is based on (21) with parameter values given in Table 2.1 for $r_\infty = 10$. The ordinate unit is $(\mu\text{mole l}^{-1})^{-1}$. (From Rubinow, 1977.).

behavior). The solid circles in Figure 2.2 represent observations (Onoue, Grossberg, Yagi & Pressman, 1968) and point up a difficulty that arises in obtaining actual data, namely, that accurate experimental measurements are very difficult to carry out at either very small or very large values of the concentration c, so that the limiting values s_0 and r_∞ cannot readily be deduced. It is clear that for c sufficiently small, measurements will ultimately strain the limits of accuracy of the measuring apparatus. The same difficulty occurs at large values of c because the measurement of the bound ligand concentration and hence r requires determining the small difference between two large quantities, namely the concentration c at the beginning and end of the equilibrium dialysis experiment.

More generally, a mathematical problem of great practical importance is to determine how the parameters appearing in $r(c)$, namely K_i and n_i for $i = 1, 2, \ldots, m$ are to be inferred from observation. This problem arises also in conjunction with the Adair equation (9), and has attracted the attention of many investigators (for example, Pauling, Pressman & Grossberg, 1944; Karush & Sonenberg, 1949; Karush, 1956; Nisonoff & Pressman, 1958; Endrenyi, Chan & Wong 1971; Werblin & Siskind, 1972). Here we shall present a simple scheme for determining these parameters from the consideration of the Scatchard plot (Rubinow, 1977).

We restrict our attention to the case $m = 2$, so that r is assumed to be of the form shown in (21). Then r and s may be written as

$$r = \frac{a_1 c + a_2 c^2}{1 + b_1 c + b_2 c^2}, \qquad s = \frac{a_1 + a_2 c}{1 + b_1 c + b_2 c^2}, \qquad (29a, b)$$

where

$$a_1 = n_1 K_1 + n_2 K_2, \qquad a_2 = (n_1 + n_2)K_1 K_2,$$
$$b_1 = K_1 + K_2, \qquad b_2 = K_1 K_2. \qquad (30)$$

We clear the denominator of (29b) and use $r = sc$, so that (29b) becomes

$$s + b_1 r + b_2 rc - a_1 - a_2 c = 0. \qquad (31)$$

The unknown coefficients above can be determined from a set of L data points (r_j, c_j), $j = 1, 2, \ldots, L$, utilizing the method of least squares. In that case the algebraic equation for the unknown coefficients are linear, so the problem is reduced to 'linear regression'. A theoretical drawback of such a procedure is that the quantity a_2/b_2, which represents the total number of binding sites per molecule $r_\infty = n_1 + n_2$, will not in general be found to be an integer.

Alternatively, suppose for the moment that the quantity r_∞ is known from the Scatchard curve. Form the function.

$$u \equiv (r_\infty - r)/s. \qquad (32)$$

By explicit calculations, we find from (29) that

$$u = \frac{r_\infty + (r_\infty b_1 - a_1)c}{a_1 + a_2 c}. \qquad (33)$$

Hence, u is precisely of the canonical form of r in (29a), with the advantage that the largest power of c appearing in either the numerator or denominator has been reduced to unity. This has the greatest significance in our example for which the largest power is 2, because we can consider u as a 'new' experimentally determined function r, and introduce a 'new Scatchard function' u/c. Thus, u and u/c are related as follows,

$$\frac{u}{c} = \frac{1}{a_1}\left[r_\infty\left(\frac{1}{c} + b_1\right) - a_1 - a_2 u\right], \qquad (34)$$

which is a linear relationship similar to the one that motivated the introduction of the Scatchard plot. Again, linear regression can be utilized to determine the unknown coefficients in the above equation.

Inasmuch as r_∞ is known only approximately from the data, we suggest that an initial 'trial' integral value of r_∞ be chosen with which to define the function u. The procedure is repeated with neighboring integral values of r_∞, and the 'best' value of r_∞ is decided upon by the criterion that the error function E of the method of least squares is minimized. Thus, corresponding to (34), the error function E is defined as

$$E = \sum_{i=1}^{L} \left(\frac{u_i}{c_i} - \frac{\alpha_1}{c_i} - \alpha_2 u_i - \beta_1 \right)^2, \tag{35}$$

where the coefficients α_1, α_2 and β_1 are to be found by the minimization criterion. Then the a_i and b_i are found by comparison of (34) and (35) from the equations

$$a_1 = \frac{r_\infty}{\alpha_1}, \qquad a_2 = -\frac{r_\infty \alpha_2}{\alpha_1},$$

$$b_1 = \frac{1}{\alpha_1}(1 + \beta_1), \qquad b_2 = -\frac{\alpha_2}{\alpha_1}, \tag{36}$$

and the n_i and K_i are in turn found from (30).

In this manner the data of Onoue *et al.* (1968) (see Figure 2.2) representing the binding of a ligand to IgM antibodies were considered, for the assumed values of r_∞ equal to 9, 10, and 11. Note that it would be very difficult to decide by extrapolation of the data points what the best value of r_∞ is, and it is useless to try to determine by extrapolation what s_0 is. The resulting parameter values of the minimization procedure are shown in Table 2.1. We note that the value $r_\infty = 10$ yields a local minimum of the

Table 2.1. *Representation of ligand binding to IgM antibodies* (*data of Onoue* et al., *1968*)

r_∞	n_1	n_2	K_1 (μmole l^{-1})$^{-1}$	K_2 (μmole l^{-1})$^{-1}$	E (μmole l^{-1})$^{-2}$
9	4.08	4.92	5.97	0.0293	0.310
10	4.07	5.93	6.24	0.0198	0.122
11	4.26	6.74	5.31	0.0131	0.219

error E and is therefore the value of choice. The solid line shown in Figure 2.2, based on (29) and the parameter values shown in the table for $r_\infty = 10$, appears to give a satisfactory fit to the data points.

Further support for the parameter choice $r_\infty = 10$ is derived from structural studies indicating that IgM is a pentameric molecule, and from the theoretical conception (Miller & Metzger, 1966) that IgM contains precisely 10 binding sites. It is likewise gratifying that the derived values of n_1 and n_2 are close to integers (see Table 2.1), as they should be on theoretical grounds. The fact that n_1 and n_2 are not both equal to 5, as might be expected, is of course unexplained by these considerations.

As a final cautionary word, it is worth remembering that the heterogeneous behavior of the macromolecular solution may be a consequence of the fact that it consists of a fraction γ of one homogeneous molecular species with n_1 binding sites, and a fraction $(1 - \gamma)$ of a second homogeneous species containing n_2 binding sites. Then r would still be represented by (21) with γn_1 replacing n_1 and $(1 - \gamma)n_1$ replacing n_2. In such a case the derived values of 'n_1' and 'n_2' would not be integers.

References

Adair, C. S. (1925). The hemoglobin system. VI. The oxygen dissociation curve of hemoglobin. *J. Biol. Chem.* **63**, 529–45.

Bjerrum, J. (1941). *Metal Ammine Formation in Aqueous Solution*, Copenhagen, P. Haase & Son.

Eadie, G. S. (1942). The inhibition of cholinesterase by physostigmine and prostigmine. *J. Biol. Chem.* **146**, 85–93.

Endrenyi, L., Chan, M.-S. & Wong, J. T.-F. (1971). Interpretation of non-hyperbolic behavior in enzymic systems. II. Quantitative characteristics of rate and binding functions. *Can. J. Biochem.* **49**, 581–98.

Karush, F. (1956). The interaction of purified antibody with optically isomeric haptens. *J. Amer. Chem. Soc.* **78**, 5519–26.

Karush, F. & Sonenberg, M. (1949). Interaction of homologous alkyl sulfates with bovine serum albumin. *J. Amer. Chem. Soc.* **71**, 1369–76.

Klotz, I. M. (1946). The application of the law of mass action to binding by proteins. Interactions with calcium. *Arch. Biochem.* **9**, 109–17.

Lineweaver, H. & Burk, D. (1934). The determination of enzyme dissociation constants. *J. Amer. Chem. Soc.* **56**, 658–66.

Miller, F. & Metzger, H. (1966). Characterization of a human macroglobulin. *J. Biol. Chem.* **241**, 1732–40.

Nisonoff, A. & Pressman, D. (1958). Heterogeneity and average combining constants of antibodies from individual rabbits. *J. Immun.* **80**, 417–28.

Onoue, K., Grossberg, A. L., Yagi, Y. & Pressman, D. (1968). Immunoglobulin M antibodies with ten combining sites. *Science* **162**, 574–6.

Pauling, L., Pressman, D. & Grossberg, A. L. (1944). The serological properties of simple substances. VII. A quantitative theory of the inhibition by haptens of the precipitation of heterogeneous antisera with antigens, and comparison with experimental results for polyhaptenic simple substances and for azoproteins. *J. Amer. Chem. Soc.* **66**, 784–92.

Rubinow, S. I. (1975). *Introduction to Mathematical Biology*, New York, Wiley. (1977). A suggested method for the resolution of Scatchard plots. *Immunochemistry* **14**, 573–6.

Scatchard, G. (1949). The attractions of proteins for small molecules and ions. *Ann. N.Y. Acad. Sci.* **51**, 658–66.

Tanford, C. (1961). *Physical Chemistry of Macromolecules*, New York, Wiley, p. 539.

Werblin, T. P. & Siskind, G. W. (1972). Distribution of antibody affinities: technique of measurement. *Immunochemistry* **9**, 987–1011.

Woolf, B. (1932). Quoted in Haldane, J. B. S. & Stern, K. G. *Allgemeine Chemie der Enzyme*, Dresden & Leipzig, Steinkopf Verlag, p. 119.

3

Allosteric and induced-fit theories of protein binding

Recall that ligand binding not obeying the hyperbolic law (2.15) is called cooperative. Many proteins display cooperative behavior, the most thoroughly studied example being hemoglobin, which possesses four binding sites for oxygen molecules. Moreover, the cooperative behavior of hemoglobin has a very important physiological significance in the transfer of oxygen from the blood of mammals to the tissue. By the same token, although the majority of enzymes obey Michaelis–Menten kinetics, a considerable number of them have been found to display cooperative behavior, and this behavior is likewise of great significance in the regulation and control of biosynthetic processes. Here we shall present the essential ideas underlying two important theoretical models of cooperative behavior of proteins: the **allosteric theory** of proteins put forward by Monod, Wyman & Changeux (1965), MWC theory for short; and the **induced-fit theory** of Koshland, Nemethy & Filmer (1966).

Allosteric theory

In MWC theory the protein is assumed to be an **oligomer**, that is to say, it is constructed from several identical subunits, or **protomers**, each of which contains one active site for binding with a ligand C. Furthermore, the subunits are assumed to be independent of each other, so that the intrinsic association constant of each and every binding site is the same. How then is cooperative behavior of the protein achieved? The answer lies in the assumption that each protomer can undergo a reversible **conformational change**, which alters its state from A to B, say. Further, the intrinsic affinity for the ligand at a binding site is different in each of the two states. Finally, it is assumed that the oligomer itself can exist in only two states, denoted by R and T, in which the protomers are either all in the A state or all in the B state.

Let $R_j(T_j)$ represent the protein state in which j ligand molecules are bound to the state R(T), in equilibrium. Then the possible reactions are expressed as

$$\left.\begin{array}{l} C + R_{j-1} \rightleftarrows R_j, \\ C + T_{j-1} \rightleftarrows T_j, \end{array}\right\} \quad j = 1, 2, 3, \ldots, n,$$

$$R_j \rightleftarrows T_j, \quad j = 0, 1, 2, \ldots, n. \tag{1}$$

The intrinsic dissociation constants in the states R and T are denoted by K_R and K_T, respectively. The dissociation constant representing the transition from R_0 to T_0 is denoted by L and called the **allosteric constant**, or more properly, the **isomerization constant**. The term 'allosteric' is meant to describe a reaction in which the binding of a ligand molecule to a protein at one site influences the binding of a second (identical or different) ligand, through the mediation of a conformation change in the protein. In practice, the term allosteric is used almost automatically to describe reactions exhibiting (positive or negative) cooperativity. Thus, as in the derivation of Adair's equation, which in fact was orignally introduced to represent the binding of oxygen to hemoglobin, the equilibria of the reactions of (1) above are represented by (see 2.6) and (2.14))

$$L = \frac{t_0}{r_0}, \quad \frac{n-j+1}{j}\frac{1}{K_R} = \frac{r_j}{cr_{j-1}},$$

$$\frac{n-j+1}{j}\frac{1}{K_T} = \frac{t_j}{ct_{j-1}}, \quad j = 1, 2, 3, \ldots, n. \tag{2}$$

Here, lower-case italic letters again represent concentrations of quantities represented by associated capital letters.

It might be supposed that n additional parameters should be introduced above to represent the last set of reactions in (1) for j equal to 1 through n, but that is not the case, because of the **principle of detailed balancing** of chemical reactions in equilibrium (Onsager, 1931). According to this principle, the frequency of transitions from one molecular state to another is equal to that in the reverse direction, when the two states are in chemical equilibrium. This has the consequence that the mathematical expression of equilibrium between any two states which are connected to each other by more than one reaction pathway is independent of the pathway. Hence, for every cyclic set of reactions, there is a **loop relation** or equation of constraint on the equilibrium constants representing the elemental reaction steps making up the cycle or 'loop' of reactions. For example, for a set of three cyclic reactions in equilibrium, for which the equilibria among the concentrations of the reactants c_j, $j = 1, 2, 3$, are represented by the association constants K_j as $c_2 = K_1 c_1$, $c_3 = K_2 c_2$, $c_3 = K_3 c_1$, then the loop relation states that $K_1 K_2 = K_3$. In the case of the equilibrium reactions of (1), there are precisely n loops, hence n equations of constraint. Thus, it turns out that no error is made in the following derivation if the last set of reactions for j equal to 1 through n is simply ignored. (For further discussion, see Rubinow, 1975, Section 2.6.)

According to (2.8) and (2.10), the saturation function is expressed as

$$Y = \frac{\sum_{j=1}^{n} j(r_j + t_j)}{n \sum_{j=0}^{n} (r_j + t_j)}. \tag{3}$$

By substituting (2) into (3), it follows with the aid of some algebraic manipulation and the use of binomial coefficients (or see Rubinow, 1975, Section 2.6) that

$$Y = \frac{Lcx(1 + cx)^{n-1} + x(1 + x)^{n-1}}{L(1 + cx)^n + (1 + x)^n}, \tag{4}$$

where

$$x = c/K_R \quad \text{and} \quad c = K_R/K_T. \tag{5}$$

This is the principal mathematical result of the MWC theory. Note that, if $L \to 0$, then (4) reduces to the noncooperative case equivalent to (2.15). We see that in addition to K_R, which only enters into the theory as a scale factor, these are two parameters, L and c. The function Y is able to represent positive cooperativity of the protein–ligand equilibrium binding behavior, but not negative cooperativity (see Chapter 4, below).

Induced-fit theory

Koshland et al. (1966) took issue with the assumption of MWC theory that the conformational change of state of the protein had to be an 'all-or-none' or 'concerted' change of state of the subunits. They proposed instead that the subunit changes its state from A to B only as a consequence of the binding of a ligand molecule (thus the description 'induced fit'). In other words, the ligand is only found bound to a subunit in the B state. Hence, proteins exist in hybrid conformational states composed of some subunits in A, and the remainder in B with attached ligand molecules.

Figure 3.1. Allowed states of a dimer according to the induced-fit theory. Subunits in the A and B states are represented by squares and circles respectively. Binding of a substrate molecule to a subunit is represented by the letter S. It is assumed that such binding to a subunit in the state A causes it to be transformed into the state B.

These remarks are illustrated in Figure 3.1, for the special case in which the protein is a dimer, i.e. it consists of two subunits. Formally, we see that the induced-fit theory is the same as the Adair theory. What has been added is the concept that the equilibrium constants attain their values as a result of conformational change of the subunits. Consequently, the equilibrium constants depend on, and can be expressed in terms of, more fundamental equilibrium constants representing the three possible kinds of pairwise subunit interactions, namely, A–B, A–A, B–B. This dependence is a function of the number of each of the different kinds of subunit interaction pairs and therefore depends on the geometry of attachment of the subunits. For example, in the case of a tetramer (oligomer of four subunits) the induced-fit theory considers three different geometrical configurations, namely, 'linear', 'square', and 'tetrahedral' arrangements of the subunits. (The first attempt to relate the equilibrium constants of the Adair theory to the geometry of the protein was made by Pauling (1935) who assumed identical subunits, and therefore introduced only one pairwise interaction constant.)

Koshland *et al.* also considered a 'concerted' model, in which all the subunits can change their state reversibly from A to B in all-or-none fashion, but only the B state can bind ligand molecules. This can be recognized as the special case of the MWC model for which $c = 0$. The state of the protein which cannot bind ligand molecules is called a **dead-end complex**. They also considered the case of non-identical subunits. For example, it is known that the bare hemoglobin molecule consists of two types of subunits.

For purposes of comparison, the states of a dimer permitted in MWC theory are illustrated in Figure 3.2. Eigen (1967) has pointed out that both the MWC theory and the induced-fit theory can both be considered special cases of a more general theory that allows hybrid conformational states, as shown in Figure 3.3. It is assumed that in a transition from one state to another, either one subunit changes its state, or a ligand molecule becomes

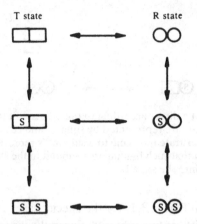

Figure 3.2. Allowed states of a dimer according to MWC theory. Only concerted changes of state of subunits are allowed, so that they are all in the state A (T state of the dimer), or in the state B (R state of the dimer).

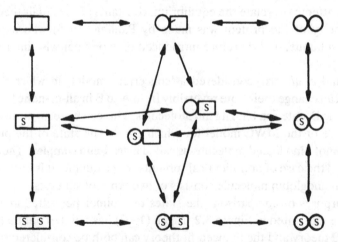

Figure 3.3. Representation of the states of a dimer in interaction with a ligand, according to the general formulation of the induced-fit model (Eigen, 1967). In each transition, either one subunit changes its state, or a ligand associates with (or dissociates from) a subunit, but not both.

bound (or unbound), but not both. The MWC model (see Figure 3.2) permits only the left-most and right-most column of states. The induced-fit model permits only those states that lie along the diagonal line running from upper left to lower right.

Actually, as stated strictly, the model requires first, ligand binding, and, secondly, transition from state A to state B, so that the allowed states lie along a 'staircase' pathway just below the diagonal line. Moreover, application of the principle of detailed balancing to one of the cycles of Figure 3.3 is equivalent to removing one of the reversible transitions that form the cycle, because no additional information is supplied to the equilibrium conditions by its inclusion. One such transition can be removed for each loop that exists in a network of reactions in equilibrium. Hence, the set of equilibrium reactions represented by Figure 3.3 is completely equivalent to the subset of reactions shown in Figure 3.4. In this form, the staircase pattern of reactions leading from upper left to lower right has been retained, in order to emphasize the formal similarity between induced-fit theory and its generalization. However, the equilibrium reaction system is also equivalent to the subset of reactions shown in Figure 3.5, chosen to emphasize the formal similarity of the general theory to MWC theory. From either Figure 3.4 or Figure 3.5, we infer that the general theory requires 9 equilibrium constants. As indicated by Eigen (1968), for the case of a tetramer the number of allowed states is 35. This number is even greater if the supposition is abandoned that the subunits in a given conformation are completely equivalent.

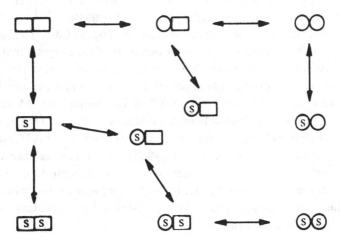

Figure 3.4. Subset of reactions equivalent to that of Figure 3.3 when the loop relations are exploited.

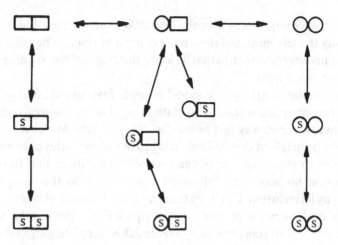

Figure 3.5. Subset of reactions equivalent to that of either Figure 3.3 or Figure 3.4.

An example: aspartate transcarbamylase

Perhaps the most extensively investigated example of a protein that was discovered to exhibit cooperative behavior (Gerhart & Pardee, 1962) is aspartate transcarbamylase (ATCase), an enzyme that catalyzes the conversion of the substrates carbamyl phosphate and aspartate to the products carbamyl aspartate and phosphate. (For reviews of its properties, see Gerhart, 1970; Jacobson & Stark, 1973; Schachman, 1974.) This conversion is the first step in the biosynthesis of pyrimidines (cytosine, thymine, and uracil), some of the fundamental building blocks of the nucleic acids. ATCase is known to react also with cytosine triphosphate (CTP) one of the end products in the biosynthetic reactions initiated by ATCase. The binding of CTP to ATCase changes the catalytic behavior of the enzyme, making it more positively cooperative in its interaction with its substrates. This effect is nullified in the presence of adenosine triphosphate (ATP), presumably as a consequence of competition with CTP at its binding site. Moreover, ATCase can be made to dissociate into two kinds of subunits. One of them, called the **catalytic subunit**, possesses the catalytic function of ATCase, but does not react with CTP. The other one, called the **regulatory subunit**, reacts with CTP and does not possess any catalytic ability. Structural studies have shown that there are six catalytic sites (one site being able to react with a pair of molecules, carbamyl phosphate and aspartate) and six regulatory sites on each native ATCase molecule.

A theoretical model of ATCase has been introduced (Dembo & Rubinow, 1977) that is capable of accounting for initial velocity studies, equilibrium binding studies of ATCase with substrate analogs that do not form products, and relaxation studies which yield quantitative information concerning changes of the enzyme molecule. No attempt is made to incorporate properties of the enzyme resulting from nucleotide (ATP, CTP) binding studies. The principal asumptions of the model are as follows.

(a) ATCase consists of three identical, noninteracting, cooperative dimers, or so-called 'allosteric units' (Markus, McClintock & Bussel, 1971).

(b) Ligand binding is highly ordered: carbamyl phosphate must bind before aspartate, and aspartate must vacate a binding site before carbamyl phosphate does.

(c) The enzyme molecule undergoes a slow concerted conformational change, such as postulated in MWC theory, but only after a carbamyl phosphate molecule is bound.

(d) The second active site is available for binding only after two conditions are fulfilled, namely, that the first site has been occupied by both a carbamyl phosphate molecule and an aspartate molecule, and that the slow conformational change of the enzyme molecule has taken place.

(e) When both sites are fully occupied, the dimer undergoes a fast conformational change, as postulated in induced-fit theory.

(f) The aspartate analog succinate can bind to the free dimeric subunit, and forms a dead-end complex unable to bind a single molecule of carbamyl phosphate.

Thus the enzyme ATCase, according to this model, exhibits some features of the induced-fit theory and some of the MWC theory. However, unlike the latter, the concerted conformational change occurs only after one of its ligands binds. Furthermore, even within a conformational state of the dimeric unit, cooperativity of ligand binding is permitted. Because all the complexity of conditions (a)–(f) above seem necessary to account for some of the ATCase data, it appears that MWC theory and induced-fit theory should be viewed as valuable simple constructs that serve to aid our understanding and thinking about the cooperative properties of proteins, even if it should turn out in the future that no proteins strictly obey these idealized models.

Crystal structure: glycogen phosphorylase

Crystallographic techniques can now be used to make direct examinations of conformational changes, typically to a resolution of about 3 Å.

An example is the work of Barford & Johnson (1989) on glycogen phosphorylase. These authors write 'Phosphorylase was one of the first enzymes whose kinetic properties were described by a two-state interpretation of the concerted allosteric model of Monod, Wyman and Changeux. Our account of the structural transition agrees closely with that envisaged in this model.'

References

Barford, D. & Johnson, L. N. (1989). The allosteric transition of glycogen phosphorylase. *Nature* **340**, 609–16.

Dembo, M. & Rubinow, S. I. (1977). A kinetic model of cooperativity in aspartate transcarbamylase. *Biophys. J.* **18**, 245–67.

Eigen, M. (1967). Kinetics of reaction control and information transfer in enzymes and nucleic acids. In *Fast Reactions and Primary Process in Chemical Kinetics*, ed. S. Claesson, Nobel Symposium vol. 5, Stockholm, Interscience, Almqvist & Wiksell, pp. 333–69.

Eigen, M. (1968). New looks and outlooks on physical enzymology. *Quart. Rev. Biophys.* **1**, 3–33.

Gerhart, J. C. (1970). A discussion of the regulatory properties of aspartate transcarbamylase from *Escherichia coli*. *Curr. Top. Cell Reg.* **2**, 275–325.

Gerhart, J. C. & Pardee, A. B. (1962). The enzymology of control by feedback inhibition. *J. Biol. Chem.* **237**, 891–6.

Jacobson, G. R. & Stark, G. R. (1973). Aspartate transcarbamylases. *Enzymes* **9**, 225–308.

Koshland, D. E. Jr, Nemethy, G. & Filmer, D. (1966). Comparison of experimental binding data and theoretical models in proteins containing subunits. *Biochemistry* **5**, 365–85.

Markus, G., McClintock, D. K. & Bussel, J. B. (1971). Conformational changes in aspartate transcarbamylase. *J. Biol. Chem.* **246**, 762–71.

Monod, J., Wyman, J. & Changeux, J. P. (1965). On the nature of allosteric transitions: a plausible model. *J. Mol. Biol.* **12**, 88–118.

Onsager, L. (1931). Reciprocal relations in irreversible processes. *Phys. Rev.* **37**, 405–26.

Pauling, L. (1935). The oxygen equilibrium of hemoglobin and its structural interpretation. *Proc. Nat. Acad. Sci., USA* **21**, 186–91.

Rubinow, S. I. (1975). *Introduction to Mathematical Biology*, New York, Wiley.

Schachman, H. K. (1974). Anatomy and physiology of a regulatory enzyme–aspartate transcarbamylase. *Harvey Lect.* **68**, 67–113.

4

Positive and negative cooperativity

The importance of cooperativity has already been mentioned at the beginning of the previous chapter, and we shall see in later chapters several specific instances where the presence of cooperativity has important physiological consequences. Typically it is the shape of the saturation function that is of physiological importance. Strictly from the point of view of physiology, it makes little difference what molecular mechanism is responsible for the observed shape. But of course a deeper understanding of the physiological behavior requires knowledge of the underlying mechanisms.

In this chapter we study more closely the concept of cooperativity that has already been introduced and discuss some of the various molecular mechanisms responsible for its appearance. We adopt here the following precise molecular definition of cooperativity. We say that an oligomer in reaction equilibrium with a ligand exhibits 'positive cooperativity' when the fraction of bound sites at any given ligand concentration is larger than that expected for the case of identical, independent sites, and 'negative cooperativity' when the fraction of bound sites is smaller than expected. Moreover, we shall see that there is an unambiguous operational realization of this definition. That is to say, the definition leads to a unique identification of cooperativity.

We have already encountered some of the special graphs – Lineweaver–Burk plot, Scatchard plot – that are employed to make easier the process of estimating the parameters that govern biochemical reactions. We shall discuss in this chapter what can be deduced about the nature of cooperativity from the appearance of these plots as well as the Hill plot.

The identification of cooperativity in binding

Let us suppose we are studying the equilibrium binding of a ligand to an oligomer of unknown character. Let the saturation function Y be given as a function of the ligand concentration x by (2.9) and (2.10), i.e. $Y = Y(x)$. We shall call the curve $Y(x)$ the **standard curve** or **standard plot**. As we have previously indicated, if the number of oligometric binding sites n is unity, or if the sites are identical and independent, then Y satisfies the equivalent of (2.15),

$$Y = Kx/(1 + Kx),\tag{1}$$

where K is the intrinsic association constant at a binding site, defined by (2.13).

For our unknown oligomer, let us examine Y for very small values of x, and determine the quantity $\bar{K}$ defined as

$$\bar{K} = (\mathrm{d}Y/\mathrm{d}x)_{x=0}.\tag{2}$$

If Y satisfied (1), then $\bar{K}$ would be equal to K. Hence, the meaning of $\bar{K}$ is that it represents the **average association constant** for the first ligand–oligomer binding process. Clearly, because $\bar{K}$ is determined from measurements of Y at very small concentrations, no higher order binding processes contribute to $\bar{K}$.

Using this quantity $\bar{K}$, we form the **reference saturation function** $\bar{Y}(x)$ defined as

$$\bar{Y} = \bar{K}x/(1 + \bar{K}x).\tag{3}$$

In accordance with our previous discussion of cooperativity, we say that the binding represented by $Y(x)$ exhibits **positive cooperativity** if for all $x > 0$,

$$Y(x) > \bar{Y}(x).\tag{4}$$

Conversely, if for all $x > 0$,

$$Y(x) < \bar{Y}(x),\tag{5}$$

then the binding is said to exhibit **negative cooperativity**.

An additional definition will prove useful. We shall say that the binding curve is **sigmoidal** if the plot of Y starts near the origin with negative curvature (concave upward) and then makes a single change in curvature, so that the curve is concave downward for sufficiently large values of ligand concentration.

The reader should be warned that our definitions do not cover all interesting possibilities, e.g. the observed binding curve could start above the reference Michaelis–Menten curve and then fall below it, or the observed curve could exhibit several changes in curvature. These possibilities are explored in the following section. Also, our definitions of cooperativity are not uniformly agreed upon in the literature. The principal reason for this disagreement is that the definitions in the literature are usually made in terms of a particular plot, while we have proceeded here from a more fundamental molecular viewpoint.

Equilibrium ligand–dimer binding

We shall confine most of our detailed study of molecular mechanisms to the dimer, because it is the simplest example of an oligomer both to analyze and to understand. First we shall assume that the sites are initially identical with respect to binding of a single ligand molecule. Thus there are only three states of the dimer to consider – when no ligand is bound, when one ligand is bound, and when two ligands are bound. We shall denote these states by C_j, where j is the number of bound ligand molecules. The directed line segments of Figure 4.1 represent the inter-state transitions, and the associated rate constants are indicated by their labels. Note in the figure that the overall rate constant for conversion from the state C_0 to the state C_1 has a factor 2, reflecting the fact that two sites are available for binding of a ligand molecule to the state C_0, and retaining for k_+ the meaning of an intrinsic forward rate constant at a single site. We denote by k_+^* and k_-^* the intrinsic forward and backward rate constants for the reaction between states C_1 and C_2; in general these constants are different from their unstarred counterparts. Comparison with the notation (2.4) used in deriving the Adair equation shows that

$$K_1 = 2K, \qquad K_2 = \tfrac{1}{2}K^*, \tag{6}$$

where $K = k_+/k_-$ and $K^* = k_+^*/k_-^*$ are the **intrinsic association constants** for binding at the first and second sites, respectively. With the present notation, the function $Y(x)$ is, from (2.9) and (2.10),

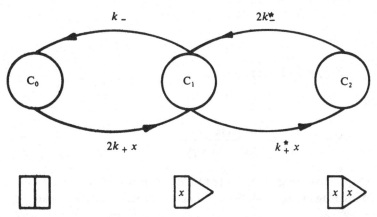

Figure 4.1. Graphical representation of transitions between the states of a dimer with initially identical sites, wherein binding of a ligand to one site changes the binding affinity of the second site. The three different states are also represented schematically, with the letter x indicating a bound ligand.

$$Y = (Kx + KK^*x^2)/(1 + 2Kx + KK^*x^2). \tag{7}$$

According to (2) and (3), the reference saturation function is

$$\bar{Y} = Kx/(1 + Kx), \tag{8}$$

and the difference $Y - \bar{Y}$ is found, with the aid of a little algebra, to satisfy

$$Y - \bar{Y} = \frac{(K^* - K)Kx^2}{(1 + Kx)(1 + 2Kx + KK^*x^2)}. \tag{9}$$

Hence, if $K^* > K$ then $Y > \bar{Y}$ and the binding is positively cooperative. Conversely, if $K^* < K$, $Y < \bar{Y}$, and the binding is negatively cooperative. In other words, binding is positively (negatively) cooperative when the intrinsic association constant for binding the second ligand molecule is greater (less) than that for the first ligand molecule. This result accords with the usual intuitive molecular meaning of positive and negative cooperativity in the literature (Cornish-Bowden & Koshland, 1975).

For our further investigations, the utilization of nondimensional variables and parameters facilitates and simplifies the mathematical investigation. Hence, we introduce the nondimensional quantities

$$\xi = Kx, \qquad \beta = K^*/K. \tag{10}$$

In terms of these, the nondimensional form of Y is

$$Y(\xi) = \frac{\xi(1 + \beta\xi)}{1 + 2\xi + \beta\xi^2}, \tag{11}$$

while

$$\bar{Y}(\xi) = \xi/(1 + \xi), \tag{12}$$

so that

$$Y - \bar{Y} = \frac{(\beta - 1)\xi^2}{(1 + \xi)(1 + 2\xi + \beta^2)}. \tag{13}$$

Hence positive (negative) cooperativity requires $\beta > 1$ ($\beta < 1$).

To investigate sigmoidality, we find by direct calculation that

$$\frac{dY}{d\xi} = \frac{1 + 2\beta\xi + \beta\xi^2}{(1 + 2\xi + \beta\xi^2)^2}, \tag{14}$$

and

$$\frac{d^2Y}{d\xi^2} = 2\frac{\beta - 2 - \beta\xi[3 + 3\beta\xi + \beta\xi^2]}{(1 + 2\xi + \beta\xi^2)^3}. \tag{15}$$

The quantity in the square brackets increases from three as ξ increases from zero. Thus the second derivative is always negtive when $\beta < 2$, but the second derivative is positive in a region of ξ near the origin when $\beta > 2$. That is, Y is sigmoidal when $\beta > 2$. This result thus shows the inappropriateness of the assumption, sometimes made, that positive cooperativity is to be identified with sigmoidality, for Y exhibits positive cooperativity but is not sigmoidal when $1 < \beta < 2$.

Figure 4.2 shows some standard curves $Y(\xi)$ for various values of β. We see that sigmoidality is apparent only for rather strong positive cooperativity ($\beta \gg 1$), while negative cooperativity never shows qualitative characteristics that are markedly different from the noncooperative case.

Various special plots are used instead of the standard plot, in order to bring out different features of the kinetics. In Chapter 1 we mentioned that the Lineweaver–Burk plot of $1/Y$ versus $1/x$ gives a straight line for noncooperative binding. In the present case we see with the definition $\eta = 1/\xi$ that

$$\frac{1}{Y} = \frac{\beta + 2\eta + \eta^2}{\beta + \eta}. \tag{16}$$

Furthermore,

$$\frac{d(1/Y)}{d\eta} = \frac{\beta + 2\beta\eta + \eta^2}{(\beta + \eta)^2}, \tag{17}$$

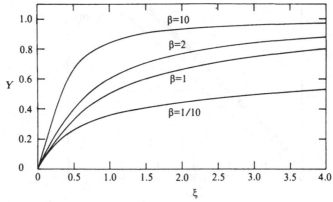

Figure 4.2. The standard plot of the nondimensional saturation function (11) for the dimer with initially identical sites.

and

$$\frac{d^2(1/Y)}{d\eta^2} = \frac{2\beta(\beta - 1)}{(\beta + \eta)^3}. \tag{18}$$

Hence, the slope of the Lineweaver–Burk curve is always positive and the curvature is positive (negative, zero) accordingly as β is greater than (less than, equal to) one. Hence the curvature of the Lineweaver–Burk plot yields a direct indication of the cooperativity of the system. Figure 4.3 displays the Lineweaver–Burk curves corresponding to Figure 4.2.

As pointed out in Chapter 2, in the Scatchard plot the function $S = Y/x$ is plotted as a function of Y, yielding a straight line for Michaelis–Menten kinetics. However, this linearity is lost when the system is cooperative. It may be readily calculated that, for $Y = Y(\xi)$ as given in (11), and $S = Y/\xi$,

$$\frac{dS}{dY} = \frac{dS}{d\xi} \cdot \frac{d\xi}{dY} = \frac{\beta - 1 - (1 + \beta\xi)^2}{1 + 2\beta\xi + \beta\xi^2}, \tag{19}$$

$$\frac{d^2S}{dY^2} = 2\beta(1 - \beta) \frac{(1 + 2\xi + \beta\xi^2)^3}{(1 + 2\beta\xi + \beta\xi^2)^3}. \tag{20}$$

Hence, the curvature of the Scatchard curve determines the cooperativity of the dimer, being always negative (positive, zero) accordingly as the coopera-

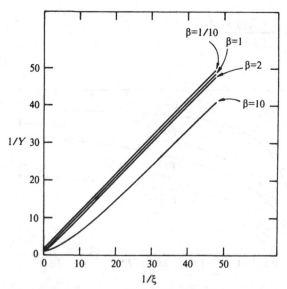

Figure 4.3. Lineweaver–Burk plots corresponding to the standard plots of Figure 4.2.

tivity is positive (negative, zero). These observations are from Endrenyi, Chan & Wong (1971). Figure 4.4 shows the family of Scatchard curves corresponding to Figure 4.2.

The Hill plot (Hill, 1910) was originally introduced as a consequence of the following type of reasoning. Suppose that an oligomer of n subunits combines with substrate molecules so rapidly that all intermediate states between the bare and the fully occupied oligomer can be neglected. Then the reaction is symbolized as

$$P_0 + nx \underset{k_{n-}}{\overset{k_{n+}}{\rightleftharpoons}} P_n, \tag{21}$$

where P_n denotes the oligomer with n bound sites. The steady state and conservation equations for such a system are

$$-k_{n+}P_0x^n + k_{n-}P_n = 0, \qquad P_0 + P_n = \bar{P}, \tag{22}$$

where $\bar{P}$ is the total oligomer concentration. The saturation function in this case is

$$Y \equiv \frac{nP_n}{n\bar{P}} = \frac{K_n x^n}{1 + K_n x^n}, \tag{23}$$

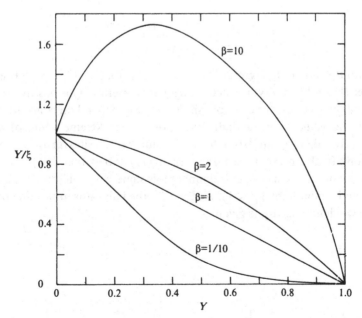

Figure 4.4. Scatchard plots corresponding to the standard plots of Figure 4.2.

where $K_n = k_{n+}/k_{n-}$. Solving for $K_n x^n$ one finds $Y/(1 - Y) = K_n x^n$, so that

$$\ln[Y/(1 - Y)] = \ln K_n + n \ln x, \tag{24}$$

and a straight line of slope n is obtained if the quantity in the left is plotted as a function of $\ln x$. Such a plot is called the **Hill plot**. The quantity

$$n_H = \frac{d \ln[Y/(1 - Y)]}{d(\ln x)} \tag{25}$$

is called the **Hill number**. A plot of the Hill number as a function of x is often routinely made, and the result used to estimate the number of binding sites on an oligomer in reaction equilibrium with a ligand. However, the above discussion shows that this interpretation is strictly valid only under very special circumstances.

In the case of a dimer obeying equation (11),

$$\ln\left(\frac{Y}{1 - Y}\right) = \ln \xi + \ln\left(\frac{1 + \beta\xi}{1 + \xi}\right), \tag{26}$$

so that

$$n_H \equiv \frac{d}{d(\ln x)} \ln\left(\frac{Y(x)}{1 - Y(x)}\right)$$

$$= \xi \frac{d}{d\xi} \ln\left(\frac{Y(\xi)}{1 - Y(\xi)}\right) = 1 + \frac{(\beta - 1)\xi}{(1 + \xi)(1 + \beta\xi)}. \tag{27}$$

Thus for the dimer, n_H is greater (less) than unity for $\xi > 0$ accordingly as β is greater (less) than unity, i.e. accordingly as cooperativity is positive (negative). In general, as this example shows, n_H will depend on concentration. Either the value of n_H at half saturation or an extreme value of n_H is frequently taken as an estimate of the number of sites involved in the reaction. In the present case analysis of (11) shows that $Y(\xi) = \frac{1}{2}$ when $\xi = \beta^{-\frac{1}{2}}$, and that Y has an extremum at this same value of ξ (a maximum if $\beta > 1$ and a minimum if $\beta < 1$). At $\xi = \beta^{-\frac{1}{2}}$, the half saturation value of n_H, which we denote as $\bar{n}_H$, is given as

$$\bar{n}_H = 1 + \frac{\beta^{\frac{1}{2}} - 1}{\beta^{\frac{1}{2}} + 1}. \tag{28}$$

Hence

$$\begin{aligned}
\bar{n}_H < 1 \quad \text{if } \beta^{\frac{1}{2}} < 1, \\
\bar{n}_H > 1 \quad \text{if } \beta^{\frac{1}{2}} > 1,
\end{aligned} \tag{29}$$

and, in fact, $\bar{n}_H \to 2$ from below if $\beta^{\frac{1}{2}} \to \infty$. The family of curves of Figure 4.2 are shown in a Hill plot in Figure 4.5.

We conclude that the Hill plot provides a sound basis for deciding on the nature of cooperativity for the dimer with identical sites. However, the maximum or half-saturation value of the Hill number has no meaning with respect to the true number of binding sites unless $\beta^{\frac{1}{2}}$ is large, when it approaches, but is less than, this number.

Let us return to the question of when the Hill kinetic representation (23) is appropriate. As we have remarked, it is certainly permissible to construct a Hill plot irrespective of the validity of (23), which was used only for motivation. But our question is important in view of the fact that many authors use the Hill law as a simple representation of cooperative kinetics.

For the dimer, the Hill law will hold if the exact kinetic expression (7) can be approximated by

$$Y = KK^*x^2/(1 + KK^*x^2). \tag{30}$$

This approximation is *permitted* if $KK^*x^2 \gg Kx$, i.e. if $K^*x \gg 1$. But the approximation is only *useful* if the concentration for half-saturation is within the permitted region $x \gg (K^*)^{-1}$. Otherwise, although (30) closely approximates the true kinetic law, the law itself for the range of x under consideration is hardly distinguishable from the description of a saturated system

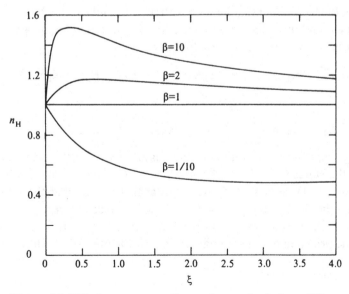

Figure 4.5. Hill plots corresponding to the standard plots of Figure 4.2.

wherein $Y \approx 1$. According to (30), half-saturation ($Y = \frac{1}{2}$) occurs when $KK^*x^2 = 1$, i.e. when $x = (KK^*)^{-\frac{1}{2}}$. This value of x will lie within the range of validity of (30) if $(KK^*)^{-\frac{1}{2}} \gg (K^*)^{-1}$, i.e. if

$$(K^*)^{\frac{1}{2}} \gg K^{\frac{1}{2}}. \tag{31}$$

Given our initial motivation, it is no surprise that a condition for the validity of the Hill approximation is that the association constant for the second binding be much larger than the corresponding constant for the first binding. (This condition also explains our observation that the half-saturation Hill number is close to the number of binding sites when $\beta \gg 1$.)

The approximation (30) will be inaccurate if x is smaller than, or comparable with, $1/K^*$. But for this range of x, the assumption $(K^*)^{\frac{1}{2}}\beta \gg K^{\frac{1}{2}}$ guarantees that Kx and therefore Y will be very small, so that large percentage errors in Y are normally immaterial. For the dimer then, we have shown that the Hill law is a useful approximation to describe binding, for all ligand concentrations, provided that the condition (31) is satisfied.

Dimer binding: nonidentical sites

We now alter our assumption that the two sites of the dimer are initially identical with respect to binding of a single ligand molecule. We label the two different sites as 1 and 2, with associated rate constants $k_{\pm 1}$ and $k_{\pm 2}$, respectively. It follows that there are now four equilibrium states of the system, the bare dimer C_0, the state with site 1 bound C_1, the state with site 2 bound C_1', and the fully bound state C_2, as illustrated in Figure 4.6.

It turns out that the saturation function is given by

$$Y = \tfrac{1}{2}(K_1x + K_2x + 2K_1K_3x^2)/(1 + K_1x + K_2x + K_1K_3x^2) \tag{32}$$

where

$$K_1 = k_{+1}/k_{-1}, \qquad K_2 = k_{+2}/k_{-2}, \qquad K_3 = k_{+3}/k_{-3},$$

and $k_{\pm 3}$ represents the rate constants associated with the transitions between the states C_1 and C_2. (When $K_3 = K_2$, (32) reduces to $\frac{1}{2}r$, where r is given by (2.21), as it should.) It might at first appear that the saturation function should depend on $K_4 = k_{+4}/k_{-4}$ as well (see Figure 4.6), but that is not the case because of the loop relation that follows from the thermodynamic principle of detailed balancing (see Chapter 3), which states that

$$K_1K_3 = K_2K_4. \tag{33}$$

Because K_1 and K_2 appear only in the combination $K_1 + K_2$, the ex-

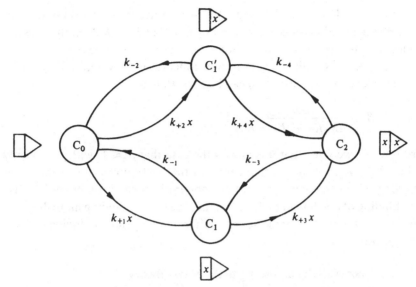

Figure 4.6. Graphical representation of transitions between the states of a dimer with two different sites. The four different states are also represented schematically, with the letter x indicating a bound ligand.

pression (32) is only a two-parameter family, in fact of the same form as (7). This is readily seen by defining

$$J = \tfrac{1}{2}(K_1 + K_2) \quad \text{and} \quad J^* = 2K_1K_3/(K_1 + K_2), \tag{34}$$

with which (32) becomes

$$Y = (Jx + JJ^*x^2)/(1 + 2Jx + JJ^*x^2). \tag{35}$$

Hence, by introduction of $\xi = Jx$, Y is expressible in the form (11) with β replaced by β', where

$$\beta' = J^*/J = 4K_1K_3/(K_1 + K_2)^2. \tag{36}$$

However, β' now has a more general significance than previously ascribed to β. For example $\beta' = 1$ implies that the geometric mean of the equilibrium constants representing the transition from C_0 to C_2 equals the arithmetic mean of the equilibrium constants for the transition from C_0 to C_1 and C_0 to C_1', i.e. from (36),

$$\sqrt{(K_1K_3)} = \tfrac{1}{2}(K_1 + K_2). \tag{37}$$

There are two important limiting cases of the system described in Figure 4.6. One occurs if $K_1 = K_2$. Then the states C_1 and C_1' are the same and the reaction scheme becomes identical with that of Figure 4.1, consisting of identical but dependent binding sites. The other important limiting case occurs if the binding sites are not identical so that $K_1 \neq K_2$, but the sites are independent, as occurs frequently for many immunoglobulins. Then $K_3 = K_2$, $K_4 = K_1$, and the saturation function is $\frac{1}{2}r$, where r is given by (2.21). More importantly, (36) can be rewritten as

$$\beta' = \frac{4K_1K_2}{4K_1K_2 + (K_1 - K_2)^2}, \tag{38}$$

from which we infer that it is always the case that $\beta' \leqslant 1$, with $\beta' = 1$ only when $K_1 = K_2$. From this there follows the well known observation that binding curves that exhibit negative cooperativity can arise either because the binding of one ligand molecule makes the second binding more difficult, or because the protein possesses two different independent binding sites for the ligand.

Cooperativity according to the MWC theory

Another mechanism for cooperativity is afforded by the allosteric theory of Monod *et al.* (see Chapter 3 above). The saturation function $Y(x)$ in this case is given by (3.4). The Michaelis constant for the first binding is obtained by examining (3.4) for small values of x. We obtain for $x \ll 1$, $Y(x) \approx (Lc + 1)x/(L + 1)$, so that our reference saturation function is

$$\bar{Y} = \frac{(Lc + 1)x}{L(1 + cx) + (1 + x)}. \tag{39}$$

Some algebra shows that

$$Y(x) - \bar{Y}(x) = \frac{L(1 - c)x[(1 + x)^{n-1} - (1 + cx)^{n-1}]}{[L(1 + cx)^n + (1 + x)^n][L(1 + cx) + 1 + x]}. \tag{40}$$

It is now apparent that (except for an isolated noncooperative case when $c = 1$, i.e. when binding to both states is the same), it is always the case that $Y(x) - \bar{Y}(x) > 0$. This analysis justifies the frequently made remark in the literature that the allosteric model can only account for positive cooperativity.

Initial velocity studies

For reactions in which an enzyme catalyzes the formation of a product, there have been extensive experimental and theoretical studies of the cooperative behavior of the enzyme as revealed by initial velocity measurements. If the enzyme is an oligomer of n subunits, it can be shown that the initial velocity of the reaction, defined as the initial rate of product formation, is in general expressible as the ratio of two n-th order polynomials. We adopt here the same general procedure for defining cooperativity with respect to initial velocity studies as we did for binding studies.

Consider for reference an 'identical site' oligomer of n subunits. Let the sites be both identical and independent (so that the intrinsic rate constants both for binding and for product formation are the same at each site). Further, assume that the value of this intrinsic product formation rate constant is such that, at large ligand concentrations, the reference oligomer achieves the same maximum initial velocity as the unknown oligomer. Cooperativity of initial velocity behavior is defined now by comparison of the unknown oligomer with the above-defined identical site oligomer, each site of which represents in some sense the average behavior of the sites of the unknown oligomer. If the initial velocity of the unknown oligomer, measured as a fraction of the maximum velocity of the reaction, is greater (less) that that of the comparison identical site oligomer, the oligomer is said to exhibit positive (negative) cooperativity.

Although we have adopted here an unambiguous operational definition of cooperativity with respect to initial velocity studies, in analogy with the analysis of binding studies, we cannot claim that this definition has a clear meaning with respect to the individual enzyme molecules. The difficulties of interpretation will become apparent in the course of our further discussion.

We shall assume that the initial velocity v approaches a finite limit V, as the substrate concentration becomes very large. We introduce the normalized initial velocity U, defined as the ratio of v to V:

$$U = v/V. \tag{41}$$

The function U has a formal resemblance to the saturation function Y. Thus, if all sites are identical and independent, then (as has been shown in (1.10)) the velocity obeys the Michaelis–Menten law

$$U = x/(K_m + x), \tag{42}$$

where K_m is the Michaelis constant.

In considering oligomeric enzymes of unknown character with normalized initial velocity functions $U(x)$, we define $\bar{K}_m$ to be the average Michaelis constant for a single ligand–oligomer reaction process, obtained from the behavior of $U(x)$ at small values of ligand concentration x:

$$\bar{K}_m = 1/[dU/dx]_{x=0}. \tag{43}$$

Note that (43) differs from the analogous equation (2) because for binding we employed the association constant while in velocity studies the Michaelis constant is primarily a measure of dissociation, although it involves the product formation process as well.

Using the $\bar{K}_m$ defined in (43) we construct the **reference initial velocity function** $\bar{U}$, based on the Michaelis–Menten law,

$$\bar{U} = x/(\bar{K}_m + x). \tag{44}$$

By analogy with our previous definition, we say that if the observed graph of U always lies above (below) the curve $\bar{U}$, then the initial velocity curve exhibits positive (negative) cooperativity.

We assume now that our enzyme is a dimer, with two identical sites, that catalyzes the conversion of the ligand into a product. All the transitions shown in Figure 4.1 are present. In addition, there is a transition from C_1 to C_0 with rate constant k_1, which transforms the ligand to a product molecule, and a transition from C_2 to C_1 with a rate constant designated by $2k_1^*$, also associated with the transformation of one of the bound ligand molecules to a product molecule. The resultant quasi-steady state initial velocity v of product formation is given as (see, for example, Rubinow, 1975, p. 67).

$$v = \frac{2e_0 x[k_1 k_m^* + k_1^* x]}{K_m K_m^* + 2K_m^* x + x^2}. \tag{45}$$

Here

$$K_m = (k_1 + k_-)/k_+, \qquad K_m^* = (k_1^* + k_-^*)/k_+^*, \tag{46}$$

and e_0 is the initial amount of enzyme present.

It is useful to introduce the nondimensional parameters and variables

$$\alpha = k_1/k_1^*, \qquad \gamma = K_m/K_m^*, \qquad V = 2e_0 k_1^*, \tag{47}$$

$$\xi = x/K_m, \qquad U = v/V.$$

With these, (45) may be written as

$$U(\xi) = \xi(\alpha + \gamma\xi)/(1 + 2\xi + \gamma\xi^2). \tag{48}$$

Note that $U \to 1$ as $\xi \to \infty$, so that U is appropriately normalized. For small ξ, $U \approx \alpha\xi$. Consequently, our definition of cooperativity requires us to contrast (48) with the reference initial velocity function

$$\bar{U} = \alpha\xi/(1 + \alpha\xi). \tag{49}$$

We find easily that

$$U(\xi) - \bar{U}(\xi) = \frac{\xi^2(\alpha^2 - 2\alpha + \gamma)}{(1 + 2\xi + \gamma\xi^2)(1 + \alpha\xi)}. \tag{50}$$

Thus, cooperativity is identified as follows:

$$\gamma > \alpha(2 - \alpha), \quad \text{positive cooperativity};$$
$$\gamma < \alpha(2 - \alpha), \quad \text{negative cooperativity.} \tag{51}$$

Hence if $\alpha > 2$, positive cooperativity is assured, but if $\alpha < 2$, cooperativity depends on the relative values of γ and α as indicated above.

Because the criterion for the identification of cooperativity involves two parameters, α and γ, which in turn are intimately related to both the binding process (ligand association and dissociation) and the product formation process (product dissociation), it should perhaps not surprise us that it seems impossible to make a relatively simple interpretation of the molecular meaning of nonzero cooperativity as identified from initial velocity studies.

To see this better, let us introduce the binding association constants $K = k_+/k_-$, and $K^* = k_+^*/k_-^*$. On the basis of our molecular understanding of the binding process alone, we expect the enzyme to display positive cooperative behavior if $K^* > K$. With $\beta = K^*/K$ (as in (10)) we see from (46) and (47) that the parameter γ is expressed as

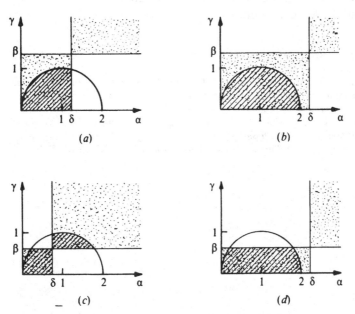

Figure 4.7. Stippled regions indicate permitted values of the parameters α and γ in (48) for the normalized initial velocity. The parts of the stippled regions that are hatched represent parameter domains corresponding to negative cooperativity. The remaining stippled regions correspond to positive cooperativity.
(a) $\beta > 1$, $\delta < 2$, (b) $\beta > 1$, $\delta > 2$, (c) $\beta < 1$, $\delta < 2$, (d) $\beta < 1$, $\delta > 2$. Recall that $\beta = K^*/K = k_+^* k_-/k_+ k_-^*$, $\alpha = k_1/k_1^*$, $\gamma = K_m/K_m^* = [(k_1 + k_-)/k_+]/[(k_1^* + k_-^*)/k_+^*]$, $\delta = k_-/k_-^*$.

$$\gamma = \beta \frac{1 + (k_1^*/k_-)\alpha}{1 + (k_1^*/k_-^*)}.$$ (52)

Note that the expression multiplying β is greater (less) than unity if α is greater (less) than k_-/k_-^*. Thus either $\alpha > \delta$ and $\gamma > \beta$ or $\alpha < \delta$ and $\gamma < \beta$, where $\delta \equiv k_-/k_-^*$. The allowed parameter space in the α–γ plane is shown as the stippled region in Figure 4.7 with the criterion of cooperativity (51) superimposed. We see from these curves that four parameters must be considered in characterizing cooperativity. Further, we see that cooperative behavior defined exclusively in terms of the binding parameter β, i.e. β greater or less than unity, does not determine cooperativity in terms of initial velocity behavior, i.e. the criterion (51).

We conclude that in spite of formal mathematical similarity between binding and velocity curves, the latter have a much more complicated molecular interpretation. The reason is that both association and dissociation processes interact in a complicated way to produce product. At the present time, no simple molecular interpretation of cooperativity in product formation has been given.

References
Cornish-Bowden, A. & Koshland, D.E. (1975). Diagnostic uses of the Hill (Logit and Nernst) plots. *J. Mol. Biol.* **95**, 201–12.
Endrenyi, L., Chan, M.-S. & Wong, J. T.-F. (1971). Interpretation of non-hyperbolic behavior in enzyme systems. II. Quantitative characteristics of rate and binding functions. *Can. J. Biochem.* **49**, 581–98.
Hill, A. V. (1910). Possible effects of the aggregation of the molecules of haemoglobin on its dissociation curves. *J. Physiol.* **40**, iv–viii.
Rubinow, S. I. (1975). *Introduction to Mathematical Biology*, New York, Wiley.

5

Graphical representations for tetramer binding

Chapter 4 provided a fairly detailed examination of the different types of cooperative behavior that are possible for a dimer. The basic mechanisms underlying the appearance of cooperative behavior are well revealed by a study of the dimer, but these mechanisms can exist in many combinations when oligomers have more than two sites. For dimers with identical sites, for example, binding of one ligand can make binding of the next easier (positive cooperativity), harder (negative cooperativity), or can have no effect (zero cooperativity). Already with trimers, the corresponding situation is much more complicated: binding of one ligand in principle can cause a conformational change that will make binding at both available sites easier, binding at one site easier and at the other harder, or binding at one site easier with the other site unaffected, etc.

In discussing n-mers, $n > 2$, an extended definition of cooperativity may be introduced with separate regions of positivity and negativity. Nonetheless the notions of cooperativity and sigmoidality are not sufficient to describe with accuracy the variety of possible behaviors of occupancy and velocity curves. Moreover, graphs of binding and velocity data can exhibit a variety of shapes, and a given shape could be due to several different underlying molecular mechanisms.

In this chapter we shall illustrate in a study of tetramers some of the issues just mentioned. The reader is referred to the work of Kenett (1978) for a much more extensive theoretical treatment.

Scatchard plots for the tetramer
The basis of our discussion is the analog of (4.11) for a tetramer

$$Y = \frac{\xi + 3a\xi^2 + 3a^2b\xi^3 + a^3b^2c\xi^4}{1 + 4\xi + a\xi^2 + 4a^2b\xi^3 + a^3b^2c\xi^4}. \tag{1}$$

Equation (1) gives the saturation function Y as a function of the dimensionless ligand concentration ξ and dimensionless ratios of association constants a, b, and c, where

$$\xi = K_1x \quad \text{and} \quad a = K_2/K_1, \quad b = K_3/K_2, \quad c = K_4/K_3. \tag{2}$$

In terms of the concentration of tetramer with i bound sites E_i ($i = 0, \ldots, 4$), the intrinsic association constants K_i are given by

$$K_1 = \frac{E_1}{4E_0x}, \qquad K_2 = \frac{2}{3}\frac{E_2}{E_1x}, \qquad K_3 = \frac{3}{2}\frac{E_3}{E_2x}, \qquad K_4 = \frac{4E_4}{E_3x}. \tag{3}$$

For the bulk of the discussions in this section, the parameters a, b, and c will be selected from the set of values 10, 1, and 0.1, giving 27 different possible combinations. These specific values represent a binding at the site in question that makes the next binding respectively easier, the same, or harder. Thus we shall make the association

$$10 \rightarrow +, \quad 1 \rightarrow 0, \quad 0.1 \rightarrow -.$$

We thereby describe individual binding events as displaying, relative to the previous binding event, a degree of positive, zero, or negative cooperativity. To give an example of the notation

$$[0 + -] \quad \text{means} \quad a = 1, \quad b = 10, \quad c = 0.1.$$

Let us now consider the 27 Scatchard plots of Y/ξ versus Y, corresponding to the 27 choices of a, b, and c (Figure 5.1, (a)–(d)). We shall analyze the results, stressing properties that allow one to distinguish at least to some extent the various sets of kinetic constants from qualitative observations of the nature of the curves.

Initial slope

The magnitude of the initial slope depends principally on the value of the parameter $a \equiv K_2/K_1$. In case $a = 1$, $b \equiv K_3/K_2$ becomes important. The situation is summarized in Table 5.1 (see p. 49). Note that if the initial slope is positive it is an increasing function of b, while initial negative slopes are independent of b. Initial slopes are uninfluenced by c.

Final slope

The value of the slope as $Y \rightarrow 1$, i.e. as $\xi \rightarrow \infty$, depends entirely on the product abc. Indeed a calculation shows that

$$\lim_{\substack{Y \rightarrow 1 \\ \text{(i.e. } \xi \rightarrow \infty)}} \frac{\mathrm{d}(Y/\xi)}{\mathrm{d}Y} = -abc. \tag{4}$$

This result can be verified from the graphs, for example the curves $[+\ 0\ 0]$ and $[+\ +\ -]$ in Figure 5.1a have the same final slope.

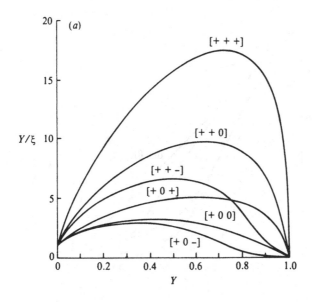

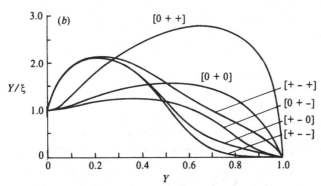

Figure 5.1. Scatchard plots for tetramers. Saturation function Y of (1) as a function of the dimensionless concentration $\xi \equiv K_1 x$. The curves are grouped into four families:
(a) $[+ + \cdot]$, $[+ 0 \cdot]$; (b) $[+ - \cdot]$, $[0 + \cdot]$; (c) $[0\ 0\ \cdot]$, $[0 - \cdot]$; (d) $[- + \cdot]$, $[- 0 \cdot)$, $[- - \cdot]$.

There are 27 curves corresponding to all possible permutations of the binding coefficient ratios K_2/K_1, K_3/K_2, K_4/K_3 chosen from the values 10, 1, or 0.1. These values are symbolized by + (positive cooperativity), 0 (noncooperativity), − (negative cooperativity). The symbol · stands for all three alternatives.

Curvature

Most situations are summarized in Table 5.2 (p. 49) with a being the dominant influence on initial curvature and c on final curvature. The three curves $[- - \cdot]$ (where · denotes +, −, or 0) constitute an exceptional case in

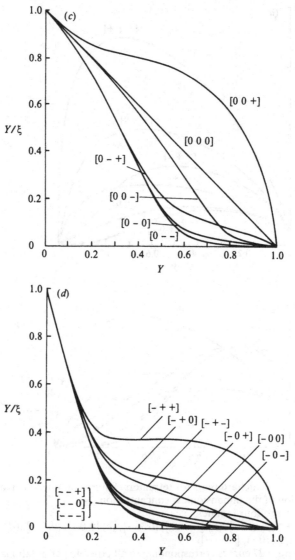

Figure 5.1. Continued.

that the final curvature is essentially zero, whatever the value of c. The zero cooperativity case [0 0 0] is a straight line devoid of curvature.

Position of maximum

Only curves of Figure 5.1a, b that start with a non-negative initial slope possess a maximum. The location of the maximum is similar for curves

Table 5.1. *Initial slopes*

Slope	Curves
Slightly positive	[0 + ·]
Definitely positive	[+ · ·]
Approximately −1	[0 0 ·], [0 − ·]
Less than −1	[− · ·]

Note: The symbol · is used whenever any one of the three symbols +, −, 0 is appropriate.

Table 5.2. *Curvatures at the boundaries of the Scatchard plots*

Boundary	Curvature	Curves
Left ($Y \to 0$)	Downwards	[+ · ·] [0 0 ·] [0 − ·]
	Upwards	[0 + ·] [− · ·]
Right ($Y \to 1$)	Downwards	[· · +]
	Upwards	[· · −]
	None	[− − ·]

with similar values of the product *abc*. The three curves [+ − ·] form an exception for which the location of the maximum is independent of *c*.

Variation of ξ

We define ξ_L as the value of ξ necessary to reach $Y = 0.9$. Table 5.3 shows how this saturation concentration depends on *a*, *b*, and *c*. The product *abc* is the most important determining factor. When $c = 1$, *ab* is also important.

We emphasize that all the above discussion applies to a special subset of parameter values for which *a*, *b*, and *c* are selected from the restricted set 10, 1, 0.1. These values are in general representative, but there are some

Table 5.3. *Values of ξ_L where $Y(\xi_L) = 0.9$*

$abc = 10^3$ $abc = 10^2$	$\begin{cases} abc = 10 \\ ab \neq 10 \end{cases}$	$abc = 1$ $ab \neq 1$	$abc = 0.1$ $ab \neq 0.1$	
				$abc = 10^{-3}$
$\begin{cases} abc = 10 \\ ab = 10 \end{cases}$	$\begin{cases} abc = 1 \\ ab = 1 \end{cases}$	$abc = 0.1$ $ab = 0.1$	$abc = 10^{-2}$	
$\xi_L: 0.06 \leqslant \xi_L \leqslant 1$	$1 \leqslant \xi_L \leqslant 10$	$10 \leqslant \xi_L \leqslant 10^2$	$10^2 \leqslant \xi_L \leqslant 10^3$	$\xi_L \simeq 6 \cdot 10^3$

features of the curves that do depend rather sensitively on the values of a, b, and c. For example, for any curve whose initial slope is not large in magnitude, one would anticipate the possibility that a small change in coefficients could have a rather large effect. This is illustrated in the case $[0 + \cdot]$ of Figure 5.1b. If the value of b is changed from $b = 10$ to $b = 2$ then the Scatchard plot changes so that a minimum appears near the origin.

A paper by Gibson & Levin (1977) mentions a minimum like the one just described and presents data that support the existence of this minimum in the Scatchard plot for [^{3}H]acetylcholine. Gibson & Levin point out that such a minimum can be found according to the concerted model for allosteric enzymes, but not for the induced-fit model. Both these models are special cases of the Adair model (Edelstein, 1975) when the coefficients of (1) respectively have the forms.

$$a = (1 + Lc^2)(1 + L)/(1 + Lc)^2,$$

$$b = (1 + Lc^3)(1 + Lc)/(1 + Lc^2)^2,$$

$$c = (1 + Lc^4)(1 + Lc^2)/(1 + Lc^3)^2,$$

and $a = b = c = K_{BB}/(K_{AB})^2$ where L, c, K_{BB}, and K_{AB} refer respectively to the constants involved in the allosteric and induced-fit models (see Chapter 3). Of course, one cannot rule out other possible special cases of Adair kinetics that could exhibit such a minimum.

Hill plot and the Hill approximation for tetramers

We call attention to a valuable discussion of Hill plots, and discuss briefly the validity of the Hill approximation for tetramers. Cornish-Bowden & Koshland (1975) displayed graphs of saturation functions for essentially the same parameter sets used here, except that they did not use K_1 to nondimensionalize ligand concentration but rather chose K_1 to have values between $10^{-1.5}$ and $10^{1.5}$ in order to center the curves around the origin. These graphs are shown in Figure 5.2. Cornish-Bowden & Koshland's careful discussion of the curves includes such matters as the uses of the asymptotes in estimating K_1 and K_4 and the effects of random and systematic error. Previously we calculated the conditions required to assure that the Hill approximation will hold for the dimer. For the tetramer, the argument is analogous to that used before. Thus it is permitted to use the saturation function.

$$Y = K_1K_2K_3K_4x^4/(1 + K_1K_2K_3K_4x^4) \tag{5}$$

as an approximation to the correct function (the dimensional version of (1)) if

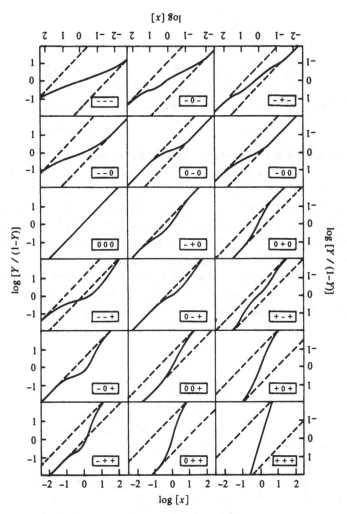

Figure 5.2. Hill plots (Figure 2 of Cornish-Bowden & Koshland, 1975). Dashed lines indicate asymptotes. Saturation function Y as a function of the dimensional concentration x. Nine new curves can be seen by viewing the figure upside down.

$$x \gg \max [K_4^{-1}, (K_3K_4)^{-\frac{1}{2}}, (K_2K_3K_4)^{-\frac{1}{3}}].$$

This approximation is useful only if the half-saturation concentration $x = (K_1K_2K_3K_4)^{\frac{1}{4}}$ lies within the permitted range of concentration. Hence the condition assuring that the Hill law may be used is

$$\max [K_4^{-1}, (K_3K_4)^{-\frac{1}{2}}, (K_2K_3K_4)^{-\frac{1}{3}}] \ll (K_1K_2K_3K_4)^{-\frac{1}{4}} \qquad (6)$$

i.e. if

$$\max [K_1^{\frac{1}{4}} K_2^{\frac{1}{4}} K_3^{\frac{1}{4}} K_4^{-\frac{3}{4}}, \; K_1^{\frac{1}{4}} K_2^{\frac{1}{4}} K_3^{-\frac{1}{4}} K_4^{-\frac{1}{4}}, \; K_1^{\frac{1}{4}} K_2^{-\frac{1}{4}} K_3^{-\frac{1}{4}} K_4^{-\frac{1}{4}}] \ll 1.$$

This condition requires that K_1 be sufficiently small compared to K_2, K_3, and K_4. As before, since errors at small concentrations are unimportant, it turns out that (5) may be used over the entire range of concentrations.

Standard and Lineweaver–Burk plots for tetramers

We close this section by presenting, for the same 27 cases as previously considered, tables summarizing the appearance of both standard plots (Figure 5.3) and Lineweaver–Burk plots (Figure 5.4). The philosophy of extracting information from such plots should be clear from the present discussion of Scatchard plots and from the discussion of Hill plots by Cornish-Bowden & Koshland (1975). It remains only to emphasize the remark that the different plots differently magnify some regions and compress others. If it is desired to extract maximum information from a given set of data, it is probably wise to make all four plots.

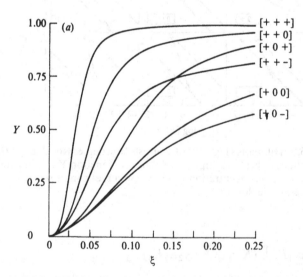

Figure 5.3. Standard plots. Saturation function Y as a function of the dimensionless concentration $\xi \equiv K_1 x$. The curves are here grouped into five families (a)–(e). The groups are the same as those in Figure 5.1 except that the family $[- - \cdot]$ appears in (e).

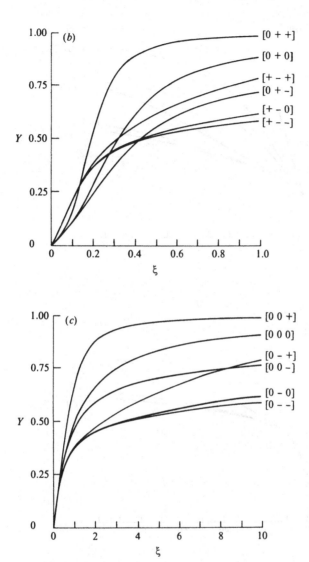

Figure 5.3. Continued.

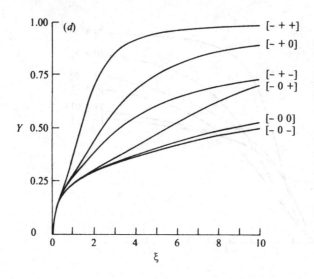

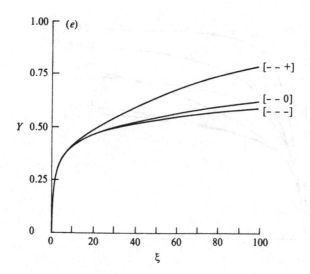

Figure 5.3. Continued.

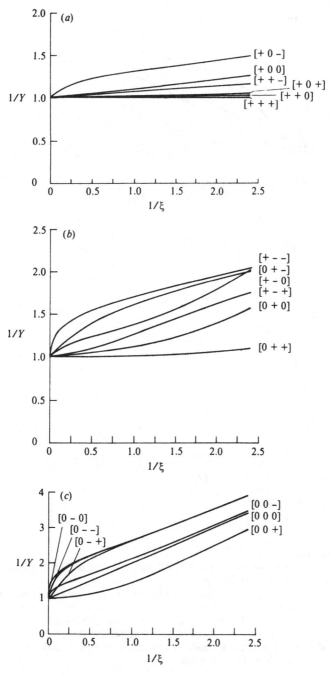

Figure 5.4. Lineweaver–Burk plots. Reciprocal of the saturation function Y as a function of the reciprocal of the dimensionless concentration ξ. The five groups (a)–(e) are as in Figure 5.3.

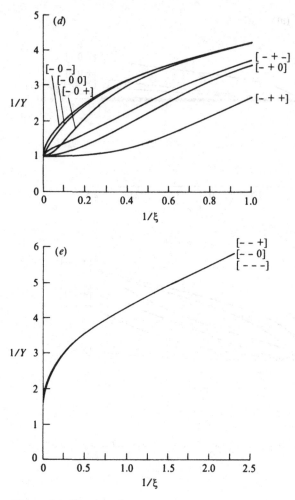

Figure 5.4. Continued.

References

Cornish-Bowden, A. & Koshland, D.E. (1975). Diagnostic uses of the Hill (Logit and Nernst) plots. *J. Mol. Biol.* **95**, 201–12.

Edelstein, S. J. (1975). Cooperative interactions of hemoglobin. *Ann. Rev. Biochem.* **44**, 209–32.

Gibson, R. & Levin, S. (1977). Distinctions between the two-state and sequential models for cooperative ligand binding. *Proc. Nat. Acad. Sci., USA* **74**, 139–43.

Kenett, R. (1978). Studies in enzyme kinetics. Ph.D. Thesis, Faculty of Math. Sci., Rehovot, Israel, The Weizmann Institute of Science.

6

Enzyme induction

Introduction

Three major modes of regulation governing the flow of metabolites through the various chemical and physical compartments of the cell are currently recognized:

(*a*) **Reversible association** of enzymes present in the cell with effector molecules arriving from outside or formed during metabolism. This mode of regulation, often accompanied by an allosteric change in the regulated enzyme, was discussed in previous chapters.

(*b*) **Covalent modification** of enzymes already present by specific kinases, phosphatases, adenylating enzymes, etc. A quantitative analysis of this mode of regulation is available (Stadtman & Chock, 1977).

(*c*) **Induction–repression**, which can be defined as an increase or decrease in the amount of an enzyme catalyzing a rate determining step(s) of a metabolic process.

Enzyme induction, in contrast to the other two modes, is a multicomponent process, because the entire protein synthesizing machinery, with its genetic control elements, is implicated. Considerable progress has been made in recent years in the elucidation of the various components participating in the induction process (Booth & Higgins, 1986; Lewin, 1987). It is becoming increasingly clear that the classical Jacob–Monod mechanism analyzed in detail below is prevalent mainly in bacterial catabolic processes. Other processes, like attenuation or transcription termination (Landick, Carey & Yanofsky, 1987; Platt, 1986), play a major role in anabolic processes. Covalent modifications, e.g. phosphorylation of regulatory proteins, have now been recognized as important control events (Weiss & Magasanik, 1988).

In this chapter we shall concentrate on the well-studied classical induction mechanisms. Examples of application of the formalism to less understood cases are discussed in the latter part of the chapter, including the perplexing phenomenon of superinduction. A procedure for determining parameters of the elusive cytoplasmic repressor, including its half-life, is presented. The reader is referred to a previous review for a more detailed analysis of the elementary systems discussed (Yagil, 1975). Several other treatments of the induction kinetics are available (Sanglier & Nicolis, 1976; Burns & Kacser, 1977; Tyson & Othmer, 1979; Berding & Harbich, 1984; Seressiotis & Bailey, 1985).

An important difference between the study of enzyme induction and the other two regulating processes is that induction is mostly studied either in the intact cell or in multicomponent, often incompletely defined, 'cell-free' systems. This reflects the fact that quite a number of sub-processes are involved, each quite complex in itself. These include DNA transcription into messenger RNA (mRNA), mRNA processing, and mRNA translation, as well as several translocatory and degradative steps. A quantitative treatment, if it is to be applicable to experimental results, requires a number of simplifying assumptions to be made in order to focus the treatment on those molecular events which are either rate limiting or subject to the regulatory interactions. We shall assume throughout that the rate of enzyme appearance is a direct measure of the amount of the principal element (active operon or analogous component) controling enzyme synthesis. We shall also assume that the intracellular effector concentration is equal, or proportional, to its extracellular concentration (this has been experimentally achieved, for instance, in the y^-, permease-less, *E. coli lac* systems or in cell-free systems). When of importance, the intracellular effector concentrations can be experimentally determined.

It is of course possible, at least in principle, to construct quantitative models which include as many steps as are known to be present. This kind of approach can be of interest in the examination of models for complex biological processes, and has been carried out in relation to problems such as cell growth and multiplication (Heinmets, 1966), as well as cellular differentiation (Edelstein, 1972).

In this discussion, the line taken is to set up a minimal number of equations required to describe the observed behavior of a system, because this seems the best way to relate to experiments. This should be borne in mind when consequences are drawn from a fit of experimental results to the prediction of a formula, or when the derived parameters are interpreted.

Effector concentration
The lactose (*lac*) operon of *E. coli*, which controls the formation of beta-galactosidase and two additional proteins, was the first genetic element

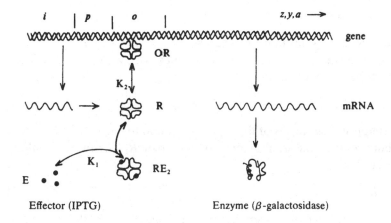

Figure 6.1. Negative induction: effect of lactose analogs on the *lac* operon.

whose regulatory circuits were elucidated, leading to many of our present concepts of regulation of enzyme synthesis (Jacob & Monod, 1961; cf. Beckwith & Zipser, 1970). The basic scheme for the operation of the *lac* operon is shown in Figure 6.1.

It is clear today that the full mechanism of induction of the *lac* operon is more complicated than this scheme. Catabolite repression operates in parallel (Figure 6.3). Further, up to three operator regions have now been identified, DNA forming loops which enable two operators to combine with a single repressor molecule (Borowiec *et al.*, 1987; Huo, Martin & Schleif, 1988). The classical scheme depicted in Figure 6.1 can nevertheless serve as a convenient starting point for the analysis of the induction process.

We shall examine the effect of the nonmetabolizable effector isopropyl-thiogalactoside (IPTG) as a basic example for evaluating the connection between effector concentration and the amount of enzyme formed in a system.

Two equilibrium relations govern the availability of the operator region for initiation by the transcribing machinery. The first equilibrium is between the protein termed the **repressor**, R, coded by the i gene, and the DNA of **free operators** O (at concentration O per unit culture volume), to form a complex OR (**blocked operator**):

$$O + R \rightleftarrows OR \qquad K_2 = O \cdot R / OR. \tag{1}$$

The **total** number of **operators** O_t in a culture population is composed of blocked and free operators:

$$O_t = O + OR. \tag{2}$$

In the second equilibrium, the repressor combines with n **effector** molecules E to form a **repressor–effector complex** RE_n with greatly reduced affinity to the operator DNA:

$$R + nE \rightleftarrows RE_n \qquad K_1 = R \cdot E^n / RE_n. \tag{3}$$

In the simplest case, forms of the repressor bound to less than n molecules contribute little to the total amount of repressor R_t , or to the resulting effects, so that

$$R_t = R + RE_n. \tag{4}$$

For a treatment where this is not the case, see Yagil & Yagil (1971). See also Chapter 4 for conditions under which the 'Hill approximation' used in (4) is valid.

We have not included OR in (4) because there are at most 4 operators per cell, i.e. $R_t \gg O_t$. Similarly, small effector molecules are, in *E. coli* at least, in large excess over the protein repressor, so that E represents practically total effector concentration ($E = E_t$). Combining (3) with (4), we find that

$$R = R_t \cdot K_1 / (K_1 + E^n). \tag{5}$$

This can be inserted into (1) to give

$$O/OR = K_2(K_1 + E^n)/R_t K_1. \tag{6}$$

We shall further define

$$\beta = \frac{O/O_t}{1 - (O/O_t)} = \frac{O}{OR}. \tag{7}$$

β is the ratio of operons available for transcription to unavailable operons. β can be regarded as an experimental quantity in those cases where the fractional rate of enzyme formation is proportional to the fraction of available operons; this has been shown to be the case in many thoroughly studied bacterial operons. Introducing (7) into (6), we arrive at

$$\beta = (K_2/R_t) + (K_2 E^n / K_1 R_t). \tag{8}$$

When no effector is present ('basal conditions'), $E = 0$, so that

$$\beta_0 = K_2/R_t. \tag{9}$$

This leads to the final form of the **induction equation**

◆◆ $$\log[(\beta - \beta_0)/\beta_0] = n \log E - \log K_1. \tag{10}$$

This equation can be put to experimental test, because, as mentioned, β can be evaluated when we know the maximal rate of enzyme production either by saturating with effector or from constitutive strains. (For further elaboration, including the effect of cell growth, see Yagil & Yagil, 1971.) A plot of $\log[(\beta - \beta_0)/\beta_0]$ versus $\log E$ should result in a straight line, if the model shown in Figure 6.1 is a sufficient description of the system. The slope of this line, n, then gives the effective stoichiometry of repressor–effector molecule interaction. From the intercept, a value for K_1, the affinity constant of effector to repressor, can be calculated. When only $\log(\beta - \beta_0)$ is plotted versus $\log E$, the intercept will yield an accurate measure of $\beta_0 = R_t/K_2$, i.e. for the ratio of total repressor concentration to its affinity to the operator region. The induction plot thus yields three parameters of the induction system.

An example is shown in Figure 6.2. Similar straight lines are obtained with many induction experiments in *E. coli*. In the *lac* system, quite a number of strains yield n values of $n = 2 \pm 0.2$ (Yagil & Yagil, 1971). This means that it is sufficient for two IPTG molecules to combine with the tetrameric repressor to reduce drastically its affinity to the operator. This is an example of half-site reactivity phenomenon proposed by Levitzki & Koshland (1975). The values of K_1 obtained in the figures are 39 and 44 $(\mu\text{mole l}^{-1})^2$ respectively. The reader is encouraged to verify this from the data in the figure.

Positive induction

We have seen that the course of induction of the *lac* operon by IPTG and by other *lac* inducers conforms closely to the predictions derived from the simple Jacob–Monod regulatory circuit. The quantitative agreement can, however, be taken as a support of the model only if other circuits lead to different types of induction plots, and this is unfortunately not always the case. Let us consider a case of positive induction, i.e. where the combination of repressor with effector converts it to a form which does associate with the operator to facilitate transcription.

In Figure 6.3, the action of cyclic AMP (cAMP) on the *lac* gene is depicted as an example of positive induction. The catabolite repression protein (CRP or CAP) has to be physically associated with the promoter region to facilitate transcription and it does so preferentially when united with effector, cAMP. In other words, a ternary complex, to be termed ORE_n, has to be formed:

$$O + RE_n \underset{}{\overset{K_2'}{\rightleftharpoons}} ORE_n \qquad K_2' = O \cdot RE_n/ORE_n. \tag{11}$$

Here,

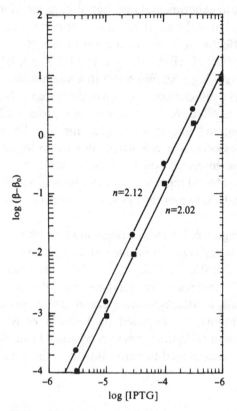

Figure 6.2. Induction of β-galactosidase in *E. coli* by IPTG. Data of Sadler & Novick (1965) are plotted according to (10). (●), Diploid strain W14D; (■), haploid strain W14. Values of $\log \beta_0 = -2.96$; 3.33 were subtracted from the ordinate; these differ by $\sim\log 2$ in accordance with the different gene dosage, and cause the two lines to be displaced by a similar factor.

$$RE_n = R_t \cdot E^n / (K_1 + E^n). \tag{12}$$

There are two problems with positive induction not encountered in the negative case. The first is that maximal induction may not be experimentally achieved because, even with maximal amounts of effector, effector concentration may not be sufficient to cause association of all R with the operator. The maximal amount of active operative can be readily seen to be:

$$ORE_n^\infty = O_t \cdot R_t / (K_2' + R_t). \tag{13}$$

Maximal rate of synthesis can be achieved only in cells in which $R_t \gg K_2'$. We shall, therefore, in the simple case, express the ratio of transcribed to nontranscribed operon in the form of

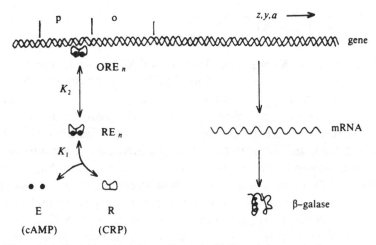

Figure 6.3. Positive induction: effect of cAMP and the catabolite repression protein (CRP) on the *lac* operon.

$$\beta' = \frac{ORE_n/ORE_n^\infty}{1 - (ORE_n/ORE_n^\infty)} \, , \tag{14}$$

and this leads to

◆◆ $\log \beta' = n \log E - \log (K_2' + R_t)/K_1 K_2,$ (15)

which is the elementary induction equation for positive induction.

The other problem with positive induction is that no enzyme should be formed when no effector is present ($\beta' = 0$ if $ORE_n = 0$). Nevertheless, in all enzyme systems examined, a basal rate of enzyme synthesis is encountered. One way to account for this is to assume that transcription of the operon can proceed to a limited extent, μ, without any repressor attached, or to an extent λ when associated with effector-free repressor. The **fraction of operons available for transcription α'** is then

$$\alpha' = (ORE_n + \lambda OR + \mu O)/(ORE_n^\infty + \mu O^\infty), \tag{16}$$

and β' is now $\alpha'/(1 - \alpha')$. This can be developed, using (6) and (11)–(14), to give a final expression

$$\log \left(\frac{\beta' - \beta_0'}{1 + \beta_0'} \right) = n \log E - \log K_{app}, \tag{17}$$

with

$$\log K_{app} = K_1 \frac{K_2'(K_2 + R_t)}{(K_2' + R_t)K_2}. \tag{18}$$

K_{app} is equal to K_1 only when both K_2 and K_2' are considerably larger than R_t. Equation (17) indicates that $\log(\beta' - \beta_0')$ should again be a linear function of the logarithm of effector concentration. This can be seen to be the case for cAMP induction of the *lac* gene in Figure 10 of Yagil (1975).

A further example of induction plots is given in Figure 6.4. Here, the data of Whitlock and Gelboin (1974) on the induction of a mammalian liver enzyme, arylhydrocarbon hydroxylase, are plotted according to (10), as a function of concentration of two of its inducers. It is seen that a straight line results. With one inducer, benzanthracene, the slope is quite close to unity, which means that a single benzanthracene molecule is sufficient to convert the controlling element into the form which facilitates enzyme formation.

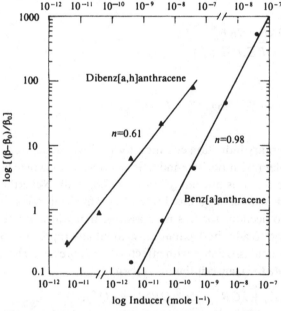

Figure 6.4. Induction of arylhydrocarbon(benzo[a]pyrene)hydroxylase by benz[a]anthracene (●) and dibenz[a,h]anthracene (▲) in cultured liver (BRL) cells. Data from Figure 3 of Whitlock & Gelboin (1974) are plotted according to (10), with $\beta_0 = 0.07, 0.15$, respectively. From the lines, values of $K_1 = 0.076$ nmol l^{-1} for benzanthracene and of 0.024 nmol l^{-1} for dibenzanthracene are calculated. To evaluate according to (17), these values have to be divided by $(1 + \beta_0)/\beta_0$, leading to $K_{app} = 1.17, 0.16$ nmol l^{-1}, respectively.

The line with a slope of less than unity observed with dibenzanthracene might be due to the very low effector concentrations involved, in which case effector may not be in excess of its receptor, as we have assumed.

We have no evidence yet, genetic or other, to establish whether a negative or positive type of control mechanism operates in the induction of any mammalian enzyme. Comparing (10) with (17) makes it apparent that they differ by the constant factor $\log [\beta_0'/(\beta_0' + 1)]$ (β' values were calculated using plateau enzyme values reached at high effector concentration; β' is, therefore, numerically equal to β, once positive induction is assumed). The linearity of the plots of Figure 6.4 thus does not permit us to determine whether induction is of the positive or negative class. It is not even possible to conclude that the controlled element is indeed the structural gene. This element may also be a stable cytoplasmic messenger ('translation control'), as will be discussed in a following section.

The effect of time

Data from the arylhydrocarbon hydroxylase system can also be utilized to illustrate the time dependence of enzyme induction. In Figure 6.5 (left-hand panels), the experimental data describing the time course of the induction are shown. Arylhydrocarbon hydroxylase, like most mammalian proteins, undergoes continuous synthesis and degradation ('turnover'). The actually observed levels represent a steady state defined by the rate of formation, expressed in terms of **translation frequency** f_2 (Yagil, 1975) and by the rate of enzyme degradation (e.g. random, with **rate constant** k_2). We shall discuss here the case when the message coding for the **enzyme** (p) neither degrades nor accumulates, or when its formation is tightly coupled to the rate of protein synthesis. In both cases, the concentration of **messenger-RNA**, M, is constant, and:

$$\dot{p} = f_2 M - k_2 p. \tag{19}$$

The steady state concentration of the induced enzyme protein, p_∞, is then

$$p_\infty = f_2 M / k_2. \tag{20}$$

Integration of this equation leads (cf. Yagil, 1975, Equation 27) to

◆◆ $(p_\infty - p)/(p_\infty - p_0) = \exp(-k_2 t). \tag{21}$

(p_0 is the initial concentration of enzyme protein.) In case of repression (i.e. a signal which either lowers f_2 or M, or increases k_2), p_0 will be larger than p_∞, as in panels c, d in Figure 6.5.

From the right-hand panels of Figure 6.5, it can be seen that a straight line

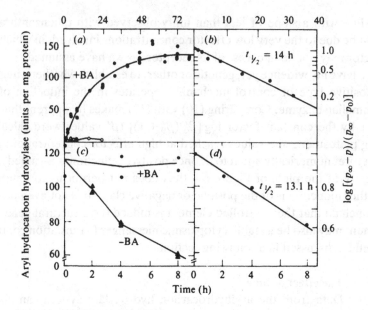

Figure 6.5. Time course of arylhydrocarbon hydroxylase induction by benz[a]anthracene (BA) in BRL cells. (*a*), (*b*), The accumulation phase. The experimental data of Whitlock & Gelboin (1973), Figure 1, are shown in panel (*a*) and replotted in panel (*b*) according to (21). (*c*), (*d*), The degradation phase. The data of Whitlock & Gelboin (1974), Figure 2, are shown in panel (*c*) and are replotted in panel (*d*) according to (2). Note the closeness of the values of $t_{\frac{1}{2}}$ obtained.

results when the logarithm of the left-hand expression of (21) is plotted against the time. The half-times evaluated from the two plots, one representing induction following addition of inducer, the other degradation following withdrawal of inducer (or addition of an inhibitor which sets $f_2 = 0$), are indeed quite close. This demonstrates that the same mechanism operates in the rate-determining step of both the inductive and the repressive phases and brings forth an interesting property of (21), namely that the time course of induction is determined by the *degradation* constant k_2 and, while the formation constant f_2 enters only implicitly, via p_∞. In other words, the kinetics of enzyme formation, like those of its disappearance, are determined solely by its degradation constant; the rate of synthesis is manifested in the level ultimately obtained. This was pointed out by Schimke (1969), as well as by Segal & Kim (1963), and serves as a routine basis for determination of enzyme 'half-lives' *in vivo* (Goldberg & St John, 1976). Equation (21) has been found to be useful in the analysis of quite a number of inducible liver enzymes, particularly if allowance for an initial latent (lag) period is given. The next stage of complication comes when active messenger is not constant as in the case described, or when enzyme is not continuously turning over. The relevant equations can be set up with ease, although each experimental situation may have its own peculiarities. Several examples are discussed elsewhere (Yagil, 1975).

Superinduction

The term superinduction has been coined to describe situations where an inhibitor of transcription or translation, instead of depressing enzyme activity, causes it to rise above the levels attained without inhibitor. This effect was first discovered when actinomycin D was added at a late stage of infection of HeLa cells with pox virus, resulting in increased accumulation of the virus coded enzyme thymidine kinase (McAuslan, 1963). It has turned out that superinduction is quite a common phenomenon (Steinberg, Levinson & Tomkins, 1975; Sen & Baltimore, 1986; Lebendiker *et al.*, 1987), and it has been shown in many cases (Kessler-Icekson & Yaffe, 1977) to be accompanied by an increase of mRNA translatable in 'cell-free' systems. Tomkins and coworkers (1969) proposed a model which attributes the superinduction of tyrosine aminotransferase in liver cells to the action of a cytoplasmic repressor capable of transforming active messenger into a form inactive in translation. The repressor and the messenger coding for it (if it is a protein) have to be much less stable than the messenger coding for the enzyme, so that when transcription inhibitor is added, this message and its products disappear rapidly to permit the message for the enzyme to be converted into its active, enzyme-producing form.

A more complicated case of superinduction has been observed in the arylhydrocarbon hydroxylase system discussed above. Whitlock & Gelboin (1973, 1974) observed that when cycloheximide is added to the system and then removed 4 h later, arylhydrocarbon hydroxylase begins to appear even when no benzanthracene or analog is added. This phenomenon occurs even when actinomycin D is added upon removal of cycloheximide, which indicates that the messenger for the enzyme can survive the cycloheximide treatment. The amount of enzyme ultimately formed was found by Whitlock & Gelboin (1974) to depend on amount of cycloheximide given. When these data are plotted according to (10), the line shown in Figure 6.6 is obtained. This line may be interpreted as meaning that cycloheximide is merely another effector of the system. However, a second translation inhibitor, puromycin, which is structurally very different from cycloheximide, has a similar superinducing effect. Further, the enzyme increase is observed quite a while after the drug is removed from the system.

We shall now show that the translation control model, depicted in Figure 6.7, which in many features is similar to the Tomkins *et al.* model (1969), can

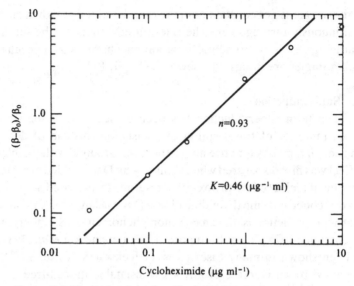

Figure 6.6. The induction of arylhydrocarbon hydroxylase by cycloheximide. The data of Whitlock & Gelboin (1974), Figure 6, are plotted according to (10).

account for most features observed by Whitlock & Gelboin in this system. In this model, it is assumed that a **translation factor** α (possibly a ribosomal protein) is required for translation to proceed. α can associate reversibly with cycloheximide (C) to form an αC **complex** which is inactive in translation:

$$\alpha + C \rightleftarrows \alpha C, \qquad K_C = \alpha \cdot C / \alpha C. \tag{22}$$

The **cytoplasmic repressor**, X, rapidly reaches a steady state due to its high **degradation constant** k_x. The steady state concentration is, according to (20),

$$X = f_X \cdot M_X / k_X. \tag{23}$$

f_X is the frequency of translation of M_X, the **messenger coding for X**. The relationship of the steady state concentration of X in presence of an inhibitor, to its concentration X_0 in absence of inhibitor will then be $[f_X/f_X^0 = C(C + \alpha C)]$

$$X = X_0 K_C / (K_C + C). \tag{24}$$

A second equilibrium exists between the **active** form M and the **inactive** form MX of the **messenger coding for the enzyme**:

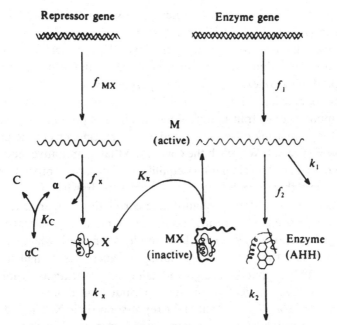

Figure 6.7. A model for superinduction by cycloheximide.
Cycloheximide associates with a translation factor, α, to form an inactive complex αC. This shuts off the synthesis of cytoplasmic repressor X, and the remaining repressor decays exponentially with a rate constant k_X. The messenger is transformed from its inactive complex with X, MX, into its active form, M (25), and begins to be translated into enzyme at a frequency f_2.

$$M + X \rightleftharpoons MX, \qquad K_X = M \cdot X/MX. \tag{25}$$

This leads to the fractional concentration of active message $(M_t = M + MX)$:

$$M/M_t = K_X/(K_X + X). \tag{26}$$

Eventually, enzyme concentration will also reach a steady state:

$$E = \frac{f_2 M}{k_2} = \frac{f_2 M_t K_X}{k_2(K_X + X)} = \frac{f_2 M_t K_X(K_C + C)}{k_2(K_C(K_X + X_0) + K_X C)}, \tag{27}$$

$(f_2 =$ frequency of translation of enzyme). This can be shown to lead to the by now familiar expression (with $n = 1$).

◆◆ $\log[(\beta - \beta_0)/\beta_0] = \log C - \log K_C.$ (28)

β is operationally taken as the ratio of enzyme activity at a certain cycloheximid concentration to the extent to which activity deviates from full induced activity: $\beta = (E - E_0)/(E_\infty - E)$, and $\beta_0 = E_0$.

The straight line we have seen in Figure 6.6 is, therefore, no less in accord with the superinduction model in Figure 6.7 than with the straight inducer role assigned to cycloheximide. The plot permits, however, the evaluation of the affinity constant of cycloheximide to the protein it associates with, whether apoinducer or translation inhibitor. The plot produces a quantitative basis for comparing efficiency of inducers – carcinogens in the present case – as in Figure 6.4. We reach the conclusion that quantitative models for superinduction based on elementary equilibrium and kinetic procedures can be set up, but that care has to be exercised in drawing a mechanistic conclusion from the observed quantitative correlation. Superinduction is, however, the favored mechanism, because of the rapid reappearance of the enzyme ($t_{\frac{1}{2}}$ about 20 min) when cycloheximide is removed, while we know from Figure 6.5 that the enzyme disappearance has a half-time of $-\ln 0.5/k_2 = 14$ h, so that the straight induction of cycloheximide is unlikely.

The quantitative application of the superinduction model permits the evaluation of the stability constant of the repressor protein X. This is done as follows. Before cycloheximide is added, a steady state level of the repressor, X_0, is present. This level is determined by the frequency of synthesis in absence of the cycloheximide, f_X^0 and rate of degradation, k_X, of X:

$$X_0 = f_X^0 M_X/k_X. \tag{29}$$

If the degradation of X is random and complete, then

$$X = X_0 \exp(-k_X t), \tag{30}$$

where t is the **time the inhibitor is present**. As before, we shall take the amount of enzyme formed (4 h out of inhibitor) to be ($f_1 = 0$ since enzyme reappearance is in the presence of actinomycin D)

$$E = \frac{f_2 M}{k_2} = \frac{f_2 M_t K_X}{k_2(K_X + X)} = \frac{f_2 M_t K_X}{k_2(K_X + X_0 e^{-k_X t})}. \tag{31}$$

When t is sufficiently large, we have

$$E_\infty = f_2 M_t/k_2. \tag{32}$$

This will lead to

◆◆ $$\log[(E_\infty - E)/E] = \log(X_0/K_X) - k_X t. \tag{33}$$

A plot of the left hand quantity against time t in cycloheximide should yield a straight line, the slope of which gives the decay constant of the cytoplasmic repressor k_X.

The data of Whitlock & Gelboin (1974) on the effect of the time in cycloheximide *before* the enzyme is allowed to be induced in the presence or absence of actinomycin D, are plotted according to (33) in Figure 6.8. Straight lines result and lead to values for the half-life of repressor ranging

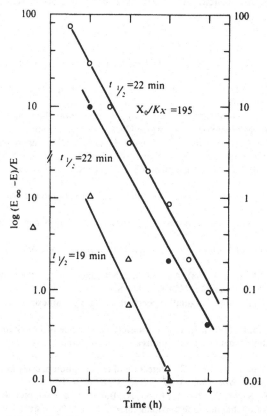

Figure 6.8. The rate of arylhydrocarbon hydroxylase reappearance after removal of cycloheximide from a culture of BRL cells. Data of Whitlock & Gelboin (1974) are plotted as function of the time the culture had been exposed to cycloheximide (i.e. the time the preexisting repressor X had been allowed to decay). (O), Data from *ibid*. Figure 6, p. 6118. Cycloheximide was followed by 4 h in actinomycin D; final enzyme levels $E_\infty = 155$ units per mg protein. ($\triangle$), Data from Figure 10, p. 6120. Benzanthracene was present together with cycloheximide, followed by 4 h in actinomycin D: $E_\infty = 540$ units per mg protein. ($\bullet$), Data from Figure 11, p. 6120; cycloheximide was followed by 4 h in actinomycin D and benzanthracene: $E_\infty = 300$ units per mg protein.

between 19 and 22 min in the three experiments evaluated. A value for X_0/K_X, the ratio of the steady state concentration of the repressor to its affinity constant to enzyme messenger, can also be obtained.

We have thus demonstrated that in this particular case a quantitative model for a superinduction can be set up and correlated with experimental data in a way which leads to conclusions concerning the quantitative behavior of the system components. This should be of considerable help in the further experimental study of the molecular details of regulation mechanisms not only in classical induction systems but also in translation controlled systems.

References

Beckwith, J. R. & Zipser, D. (1970). *The Lactose Operon*, New York, Cold Spring Harbor Publ.

Berding, C. & Harbich, T. (1984). On the dynamics of a simple biochemical control circuit. *Biol. Cybern.* **49**, 209–19.

Booth, I. R. & Higgins, C. F., eds. (1986). *Regulation of gene expression.* 39th Symposium of the Society for General Microbiology, Cambridge University Press, Cambridge, UK.

Borowiec, J. A., Zhang, L., Sasse-Dwight, S. & Gralla, J. D. (1987). DNA supercoiling promotes formation of a bent repression loop in *lac*-DNA. *J. Mol. Biol.* **196**, 101–11.

Burns, J. A. & Kacser, H. (1977). Allosteric repression, an analysis. *J. Theoret. Biol.* **68**, 19–213.

Edelstein, B. (1972). The dynamics of cellular differentiation and associated pattern formation. *J. Theoret. Biol.* **37**, 221–43.

Goldberg, A. L. & St John, A. C. (1976). Intracellular protein degradation in mammalian and bacterial cells. *Ann. Rev. Biochem.* **45**, 747–803.

Heinmets, F. (1966). *Analysis of Normal and Abnormal Cell Growth*, New York, Plenum Publ. Corp., Inc.

Huo, L., Martin, J. K. & Schleif, R. (1988). Alternative DNA loops regulate the arabinose operon in *E. coli. Proc. Natl. Acad. Sci. USA* **85**, 5444–8.

Jacob, F. & Monod, J. (1961). Genetic regulatory mechanisms in the synthesis of proteins. *J. Mol. Biol.* **3**, 318–56.

Kessler-Icekson, G. & Yaffe, D. (1977). Increased translatability in a cell-free system of RNA extracted from actinomycin D-treated cultures. *Biochem. Biophys. Res. Commun.* **75**, 62–8.

Landick, R., Carey, J. & Yanofsky, C. (1987). Detection of transcription pausing *in vivo* in the trp operon leader region. *Proc. Natl. Acad. Sci. USA* **84**, 1507–11.

Lebendiker, M. A., Tal, S., Sager, D., Pilo, S., Eilon, A., Banai, Y. & Kaempfer, R. (1987). Superinduction of the human gene encoding interferon. *EMBO J.* **6**, 585–9.

Levitzki, A. & Koshland, D. E. (1975). The role of negative cooperativity and half-of-the-sites reactivity in enzyme regulation. *Curr. Top. Cell. Reg.* **10**, 1–40.

Lewin, B. (1987). *Genes*, 3rd edition. Wiley, New York, Chapters 10, 11.

McAuslan, B. R. (1963). The induction and repression of thymidine kinase in the pox-virus infected HeLa cell. *Virology* **21**, 383–9.

Platt, T. (1986). Transcription termination and the regulation of gene expression. *Ann. Rev. Bioch.* **55**, 339–72.

Sadler, J. M. & Novick, A. (1965). The properties of repressor and the kinetics of its action. *J. Mol. Biol.* **12**, 305–27.

Sanglier, M. & Nicolis, G. (1976). Sustained oscillations and threshold phenomena in an operon control circuit. *Biophys. Chem.* **4**, 113–21.

Schimke, R. T. (1969). On the roles of synthesis and degradation in regulation of enzyme levels in mammalian tissues. *Curr. Top. Cell. Reg.* **1**, 77–124.

Segal, H. L. & Kim, Y. S. (1963). Glucocorticoid stimulation of the biosynthesis of glutamic alanine transaminase. *Proc. Nat. Acad. Sci., USA* **50**, 912–18.

Sen, R. & Baltimore, D. (1986). Inducibility of kappa immunoglobulin enhancer binding protein NK-kB by a posttranslational control mechanism. *Cell* **47**, 921–8.

Seressiotis, A. & Bailey, J. E. (1985). Intracellular equilibrium calculations based on small system thermodynamics. *Biotech. Bioeng.* **29**, 29, 1520–3.

Stadtman, E. R. & Chock, P. B. (1977). Superiority of interconnectable enzyme cascades in metabolic regulation. *Proc. Nat. Acad. Sci., USA* **74**, 2761–6.

Steinberg, R. A., Levinson, B. B. & Tomkins, G. M. (1975). 'Superinduction' of tyrosine aminotransferase by actinomycin D: a reevaluation. *Cell* **5**, 29–35.

Tomkins, G. M., Gelehrter, T. D., Granner, D., Martin, D., Samuels, H. H. & Thomson, E. B. (1969). Control of specific gene expression in higher organisms. *Science* **166**, 1474–80.

Tyson, J. J. & Othmer, H. G. (1979). The dynamics of feedback control circuits in biochemical pathways. *Prog. Theoret. Biol.* **5**, 1–49.

Weiss, V. & Magasanik, B. (1988). Phosphorylation of nitrogen regulator I (NRI) of *E. coli*. *Proc. Natl. Acad. Sci. USA* **85**, 8919–23.

Whitlock, J. P. & Gelboin, H. V. (1973). Induction of aryl hydrocarbon (benzo[a]pyrene)-hydroxylase in liver cell culture by temporary inhibition of protein synthesis. *J. Biol. Chem.* **248**, 6114–21.

Whitlock, J. P. & Gelboin, H. V. (1974). Aryl hydrocarbon (benzo[a]pyrene)hydroxylase induction in rat liver cells in culture. *J. Biol. Chem.* **249**, 2616–23.

Yagil, G. (1975). Quantitative aspects of protein induction. *Curr. Top. Cell. Reg.* **9**, 183–235.

Yagil, G. & Yagil, E. (1971). On the relation between effector concentration and the rate of induced enzyme synthesis. *Biophys. J.* **11**, 11–27.

7

Molecular models for receptor to adenylate cyclase coupling

Introduction

Numerous biochemical processes are triggered by the interaction of a specific membrane receptor with a specific ligand. In all of these cases, subsequent to the receptor-ligand binding step a post-receptor event is elicited. This post-receptor event may be either the activation of an enzyme such as adenylate cyclase or opening of a 'gate' to allow for specific ion fluxes. Little is known about the molecular events occurring subsequent to ligand binding to the receptor. Furthermore, it is not at all clear whether the receptor and the post-receptor entity responsible for generating the biochemical signal are pre-coupled to each other or are functionally uncoupled from each other and become coupled subsequent to ligand binding to the receptor. This question is the topic to which this study is devoted.

One of the means to reveal information about the state of coupling between the receptor and the enzyme is to perform a kinetic analysis of the rate of appearance of the active adenylate cyclase as a function of hormone concentration. Usually the accumulation of active cyclase is a very rapid process that is difficult to monitor. Fortunately, many adenylate cyclase systems can be activated to a permanently active state in the presence of hormone and Gpp(NH)p (guanylyl-imido-diphosphate), a non-hydrolysable analog of GTP (guanosine triphosphate) (Schramm & Rodbell, 1975; Pfeuffer & Helmreich, 1975; Sevilla et al., 1976; Levitzki, Sevilla & Steer, 1976; Spiegel et al., 1976). Although the highly active state is induced by hormone and therefore must be formed through interaction with the receptor, once formed this state is a permanent property of the enzyme and no longer involves receptor intervention.

We shall be concerned with the catecholamine-stimulated adenylate cyclase. Here propranolol, a specific β-adrenergic blocker of the catecholamine-stimulated adenylate cyclase system can no longer reduce in any way the activity of the highly active cyclase already formed but can block further activation of a partially activated enzyme (Sevilla *et al.*, 1976; Spiegel *et al.*, 1976). A detailed description of this phenomenon has already been presented (Sevilla *et al.*, 1976). Gpp(NH)p replaces the natural allosteric activator GTP. When GTP is present the active state of the enzyme is continuously reverted to its inactive state at a very rapid rate, concomitantly with the hydrolysis of GTP at the regulatory site (Cassel & Selinger, 1976). Thus under physiological conditions a steady state concentration of active enzyme is achieved (Levitzki, 1977; Sevilla & Levitzki, 1977). This steady state level is formed extremely fast and therefore the appearance of this state is impossible to monitor as a function of time. Only the steady state activity level can be measured. On the other hand, Gpp(NH)p activation is a rather slow process. Because of the irreversibility of the Gpp(NH)p activation process and its relative simplicity, it has become possible to probe into the state of enzyme–receptor coupling.

The approach used in this study was to analyze the kinetics of the process of enzyme activation induced by hormone and Gpp(NH)p in terms of different molecular models, allowing for different modes of enzyme–receptor coupling. These models also make specific predictions concerning the nature of the receptor itself, and the nature of hormone binding to the receptor. Therefore a detailed exploration of the different models of enzyme receptor coupling has been performed. Once this has been done, explicit experiments can be designed in order to test which of the different mechanisms of coupling is likely to be operating in the experimental system.

Indeed, such experiments have been carried out and are summarized at the end of this chapter. The more detailed experimental data are given elsewhere (Tolkovsky & Levitzki, 1978*a, b, c*). The theoretical analysis presented in this section is general and may be applied to receptor cyclase systems other than the β-adrenergic receptor dependent adenylate cyclase. In each of the models analyzed, expressions will be derived for (*a*) the binding of hormone, (*b*) the accumulation of the activated state of the enzyme in the presence of Gpp(NH)p, and (*c*) the steady state level of active enzyme in the presence of GTP. Also, the consequences are formulated of changing the receptor concentration or of the enzyme concentration, both on the binding of hormone and on the kinetics of enzyme activation.

Theory

The β-adrenergic-cyclase-activated system contains at least two basic components, the receptor and the adenylate cyclase. In order to elucidate the way by which the interaction of receptor and cyclase is dependent on hormone, a complete description of this system in terms of hormone–receptor–cyclase interactions must be analyzed. Boeynams & Dumont (1975) have listed several possible models for receptor–enzyme interactions and their dependence on hormone concentration in general terms.

We shall now provide kinetic models of a system consisting of the three components hormone, receptor, and enzyme (cyclase) in kinetic terms, in order to describe the adenylate cyclase in turkey red blood cells. The analysis will be divided into five parts. A general derivation of each model in terms of active cyclase will be given along with the experimental predictions associated with each model.

In all models it will be assumed that the binding of hormone to the receptor is rapid and reversible. Such rapid equilibrium is well established for the β-adrenergic receptors of turkey erythrocytes analyzed in the present study (Levitzki, Steer & Atlas, 1974; Levitzki, Sevilla, Atlas & Steer, 1975). It is well documented that the interactions between the hormone and the β-adrenergic receptor are completely reversible since the introduction of a β-adrenergic blocker completely blocks the hormone activation in a competitive fashion (Levitzki *et al.*, 1974). It is also known that one can easily wash the hormone and lose the activating effect in the absence of Gpp(NH)p.

Another major assumption is that during the process of activation either the rate of formation of the hormone–receptor–enzyme complex (HRE) is fast compared to the rate of transition to the permanently active form or the transition step is fast compared to the rate of formation of HRE. This second assumption is justified on the basis of two sets of observations. (*a*) The accumulation of cyclic AMP (cAMP) in the adenylate cyclase assay, in the presence of GTP, is linear with time. (*b*) The appearance of the permanently active state of the enzyme in the presence of Gpp(NH)p is a purely exponential process without a lag period. (If the species HRE were to accumulate to a significant fraction of the total enzyme concentration at a rate comparable to the rate of transition to permanently active form, the appearance of the permanent state would occur nonexponentially with a lag time equal to the time it takes for HRE to reach a steady state concentration.)

The different models for receptor–enzyme interactions will now be listed. Because of our assumptions it can be presumed that either the concentration *HRE* is always small compared to total enzyme E_T because of the fast transition rate or that an equilibrium level of HRE is achieved rapidly compared to the step of enzyme activation. Swillens & Dumont (1976) suggested a few models where the binding of hormone to the receptor is not a rapid equilibrium process. This particular case is not treated here.

The precoupled receptor–cyclase model (I)
This model is described by the relationship

$$H + RE \underset{k_2}{\overset{k_1}{\rightleftharpoons}} HRE \overset{k_3}{\longrightarrow} HRE', \tag{1}$$

where H is hormone, RE the receptor–enzyme complex, k_1 and k_2 are the rate constants for hormone–receptor binding, and k_3 is the rate constant characterizing the conversion of the enzyme from the inactive state HRE to the stable active state HRE'.

In the absence of guanyl nucleotide, HRE accumulates and thus represents the bound complex of hormone to the receptor. In the presence of nucleotide, HRE' is also formed. If the nucleotide is non-hydrolysable, like Gpp(NH)p, HRE' accumulates until no more free receptor is available. In the presence of a hydrolysable nucleotide like GTP, HRE' is rapidly converted back to HRE.

Hormone binding
Using (1) one can obtain an expression for the dependence of *HRE* on *H*. (We use italic to denote concentration.) Note first that the conservation equation for RE is

$$RE_T = RE + HRE. \tag{2}$$

Since $H \gg RE_T$, $H \simeq H_T$,

$$HRE = \frac{k_1 H \cdot RE}{k_2}. \tag{3}$$

From (2) and (3) it follows that

$$RE_T = RE[1 + (k_1 H/k_2)]. \tag{4}$$

Consequently

$$HRE = \frac{k_1 H \cdot RE_T}{k_2[1 + (k_1 H/k_2)]} = \frac{H \cdot RE_T}{(k_2/k_1) + H}. \tag{5}$$

The ratio k_2/k_1 is in fact the hormone–receptor dissociation constant, which will be denoted by K_H:

$$K_H = k_2/k_1. \tag{6}$$

Equation (5) therefore takes the form

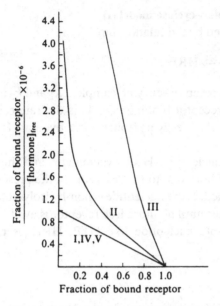

Figure 7.1. Simulation of hormone binding according to the different models of receptor to enzyme coupling. Simulation conditions presented in the figure are: $R_T = E_T = 10K_E$, $K_H = 1.0 \times 10^{-6}$ mole l^{-1}. Simulation was conducted with hormone concentrations from 1.0×10^{-8} mole l^{-1} to 1.0×10^{-5} mole l^{-1}. Changing the simulation conditions does not change the general characteristics of the curve. The Roman numerals signify the model according to which the simulation was conducted.

$$HRE = \frac{H \cdot RE_T}{K_H + H},\tag{7}$$

which represents a noncooperative binding process (Figure 7.1).

The formation of permanently active enzyme

Using (1) we now derive an expression for the rate of accumulation of the activated species of the enzyme HRE′, in the presence of Gpp(NH)p. Since HRE′ can now be formed, the conservation equation (2) is replaced by

$$RE_T = RE + HRE + HRE'.\tag{8}$$

Applying the quasi-steady state assumption (Chapter 1) for HRE we obtain

$$HRE = [k_1 H(RE_T - HRE - HRE')]/(k_2 + k_3).\tag{9}$$

Thus

$$HRE = \frac{H}{[(k_2 + k_3)/k_1] + H}(RE_T - HRE').$$ (10)

The rate of accumulation of the active species HRE' is given by

$$\frac{dHRE'}{dt} = k_3 HRE = \frac{k_3 H(RE_T - HRE')}{[(k_2 + k_3)/k_1] + H}.$$ (11)

Upon integration, taking into account that $HRE' = 0$ at $t = 0$, one obtains

$$HRE' = RE_T\left\{1 - \exp\left(-\frac{k_3 H}{[(k_2 + k_3)/k_1] + H}\right)t\right\}.$$ (12)

The steady state level of active enzyme in the presence of GTP

In the presence of GTP the scheme presented by (1) must be modified to

$$H + RE \underset{k_2}{\overset{k_1}{\rightleftarrows}} HRE \underset{k_4}{\overset{k_3}{\rightleftarrows}} HRE'$$ (13)

Here k_3 is the rate constant of enzyme activation and k_4 is the rate constant describing the conversion of the active enzyme to its inactive state, concomitantly with the hydrolysis of GTP.

Applying once again the steady state approximation for HRE we obtain

$$k_1 RE \cdot H + k_4 HRE' - (k_2 + k_3)HRE = 0.$$ (14)

Thus

$$HRE = [k_1 REH + k_4 HRE'](k_2 + k_3).$$ (15)

Consequently

$$\frac{dHRE'}{dt} = \frac{k_3 k_1 RE_T}{k_2 + k_3 + k_1 H} - HRE' \cdot \frac{k_1 H(k_3 + k_4) + k_2 k_4}{k_2 + k_3 + k_1 H}$$ (16)

Integrating (16) and applying the initial condition $HRE = 0$ when $t = 0$, one obtains

$$HRE' = \frac{[k_3/(k_3 + k_4)]H \cdot RE_T}{[k_2 k_4/k_1(k_3 + k_4)] + H}\left\{1 - \exp-\left(\frac{(k_2 k_4/k_1) + (k_3 + k_4)H}{[(k_2 + k_3)/k_1] + H}\right)t\right\}.$$ (17)

When steady state is achieved (t is large), (17) yields

$$HRE' = \frac{\{1/[1 + (k_4/k_3)]\}RE_T \cdot H}{\{k_2/[k_1(1 + (k_3/k_4))]\} + H}.$$ (18)

If $k_4 \gg k_3$ (as in the case of turkey erythrocyte adenylate cyclase (Levitzki, 1977)), the affinity towards hormone measured in the presence of GTP is identical to that measured directly by binding, i.e. $k_2/k_1 = k_H$. The maximal activity of the enzyme obtained in the presence of high hormone is given by

$$HRE'_{max} = \frac{RE_T}{1 + (k_4/k_3)} \tag{19}$$

which indicates that $HRE'_{max} \ll RE_T$ when $k_4 \gg k_3$. This expression is similar to the one obtained by Levitzki (1977). A somewhat modified model which builds a steady state level of HRE' in the presence of GTP,

$$HR + E \underset{k_2}{\overset{k_1}{\rightleftharpoons}} HRE \xrightarrow{k_3} HRE' \xrightarrow{k_4} H + RE,$$

yields an equation for HRE' which is similar to (18). The expression for HRE'_{max} is identical for the two models. The rate governing the reversal process may be experimentally derived after achieving a steady state level of HRE' by the concomitant addition of the β-adrenergic antagonist PPL and labeled ATP. HRE' will decay rapidly to the ground state by a first order process and cAMP will accumulate exponentially (Tolkovsky & Levitzki, 1978b).

This model predicts that (a) there is no basal cyclase activity in the absence of hormone; (b) the rate at which the enzyme is activated depends on hormone concentration and is saturable according to a Michaelian function; (c) the affinity of the hormone to the receptor as measured kinetically should be identical or lower than that measured directly by binding experiments; (d) the decrease of receptor concentration or of enzyme concentration by irreversible inactivation should not affect the *rate* at which the activated species HRE' appears but should decrease the maximal level of HRE' attainable either in the presence of Gpp(NH)p or in the presence of GTP.

The dissociation receptor–cyclase model (II)
The model is described by the equation:

$$RE + H \underset{K_H}{\rightleftharpoons} HRE \underset{K_E}{\rightleftharpoons} HR + E; \qquad E \to E'. \tag{20}$$

The model postulates that E' is the active species of the cyclase. As in model I, the receptor is obligatorily coupled either to E or to R.

Hormone binding
Both R and RE bind hormone. In the absence of guanyl nucleotides, E' is not formed and one can write

$$HRE = \frac{REH}{K_H},\tag{21}$$

and

$$HR = \frac{K_E HRE}{E} = \frac{K_E RE \cdot H}{K_H E}\tag{22}$$

The conservation laws are

$$E_T = RE + HRE + E\tag{23}$$

and

$$R_T = RE + HRE + HR,\tag{24}$$

from which one obtains

$$E = E_T - RE - HRE = E_T - RE\left[1 - \frac{H}{K_H}\right]$$

$$= E_T - RE\left\{\frac{K_H + H}{K_H}\right\},\tag{25}$$

$$B = HR + HRE,\tag{26}$$

where B is the concentration of the bound hormone. It follows from (22) and (23) that

$$B = \frac{RE \cdot H}{K_H}[1 + (K_E/E)].\tag{27}$$

From (25), (26) and (27) one obtains

$$B = \frac{(R_T - B)H}{K_H}\left(1 + \frac{K_E K_H}{E_T K_H - R_T(K_H + H) + B(K_H + H)}\right),\tag{28}$$

which upon rearrangement yields

$$B^2 + \frac{B}{K_H + H}\left\{K_H\left(E_T + \frac{K_E H}{K_H + H}\right) - R_T(2H + K_H)\right\}$$

$$+ \frac{(R_T)^2 H(K_H + H) - R_T \cdot H(E_T + K_E)k_H}{(K_H + H)^2} = 0.\tag{29}$$

The dependence of B on H is not immediately apparent. However, simulation of (29) reveals that the binding is negatively cooperative (Figure 7.1). The extent of negative cooperativity becomes more pronounced as K_E is reduced relative to E_T. Reducing K_E results in the relative increase of HRE, the apparently low affinity form, relative to HR, the apparently high affinity species.

The rate of formation of the permanently active enzyme

The process of enzyme activation can be described by the scheme

$$RE + H \underset{k_2}{\overset{k_1}{\rightleftharpoons}} HRE \underset{k_4}{\overset{k_3}{\rightleftharpoons}} HR + E; \qquad E \overset{k_5}{\longrightarrow} E', \tag{30}$$

where $K_H = k_2/k_1$ and $K_E = k_3/k_4$. The accumulation of E' describes the activation of the enzyme to its permanently active state.

From the relationships

$$E_T = E + E' + RE + HRE, \tag{31}$$

$$R_T = RE + HRE + HRE, \tag{32}$$

$$HRE = \frac{E \cdot HR}{K_E} = \frac{H \cdot RE}{K_H} \quad \text{and} \quad RE = \frac{E \cdot HRK_H}{K_E H}, \tag{33}$$

one can write

$$E_T - E' = E + RE + HRE = E\left(1 + \frac{HRK_H}{K_E H} + \frac{HR}{K_E}\right). \tag{34}$$

From (34) it follows that:

$$HR = R_T - (E_T - E') + E. \tag{35}$$

Inserting (35) into (34) one obtains:

$$E^2 + E\left(\frac{K_E H}{K_H + H} + R_T - E_T + E'\right) - \frac{(E_T - E' K_E H)}{k_H + H} = 0. \tag{36}$$

When $k_5 > k_4$,

$$E \ll E' + RE + HRE, \tag{37}$$

and (36) simplifies to:

$$E = \frac{(E_T - E')(k_3/k_4)H}{(k_3/k_4)H + [R_T - (E_T - E')](K_H + H)}. \tag{38}$$

The rate of formation of E' is given by [see (30)]

$$\frac{dE'}{dt} = k_5 E. \tag{39}$$

Inserting (38) into (39) one obtains

$$\frac{dE'}{dt} = \frac{(k_5 k_3/k_4)(E_T - E')H}{(k_3/k_4)H + (R_T - (E_T - E')(K_H + H))},$$ (40)

which can be rewritten as

$$\frac{dE'}{dt} = \frac{(E_T - E')}{k_5 + [k_4(K_H + H)/k_5 k_3 H][R_T - E_T - E')]},$$ (41)

or

$$\frac{dE'}{dt} = \frac{(E_T - E')}{k_5 + [k_4(K_H + H)/k_5 k_3 H]R_T - [k_4(K_H + H)/k_5 k_3 H](E_T - E')}.$$ (42)

Let us define

$$a = k_5 + \frac{k_4}{k_5 k_3} \cdot \frac{K_H + H}{H}(R_T - E_T),$$ (43)

$$b = \frac{k_4}{k_5 k_3} \cdot \frac{K_H + H}{H}.$$ (44)

From equations (42), (43) and (44) one obtains

$$\frac{dE'}{dt} = \frac{E_T - E'}{a + bE'}.$$ (45)

With these (42) takes the form

$$\frac{dE'}{(E_T/a) - (E'/a)} + \frac{E' \, dE'}{(E_T/b) - (E'/b)} = dt.$$ (46)

Upon integration one obtains the expression

$$\frac{\ln E_T}{E_T - E'} - \frac{bE'}{a + b^2 E_T} = \frac{t}{a + b^2 E_T}.$$ (47)

From (47) it is apparent that the kinetics of accumulation of E' deviates from first-order kinetics. It can also be seen that reducing R_T or E_T by irreversible inactivation results in both the reduction of the maximal level of E' attainable and a decrease in its rate of formation.

To summarize, the predictions of the dissociation model are as follows: (a) In the absence of hormone there might be basal activity; if not, all enzyme units are coupled to receptor units. (b) The binding of hormone will be negatively cooperative although one intrinsic binding constant governs the hormone binding. (This is because there are two hormone-bound species present in parallel, HR and HRE.) (*c*) The kinetics which govern the accumulation of E' are not first order. (*d*) The effect of irreversibly reducing either R_T or E_T will slow down the rate of accumulation of E' and decrease the maximal level of E' attained.

The floating receptor–enzyme equilibrium model (III)

In this model it is assumed that equilibrium exists between the hormone–receptor complex and the enzyme. The scheme is:

$$H + R \underset{k_2}{\overset{k_1}{\rightleftharpoons}} HR; \qquad HR + E \underset{k_4}{\overset{k_3}{\rightleftharpoons}} HRE \overset{k_5}{\longrightarrow} HRE', \tag{48}$$

where $K_H = k_2/k_1$ and $K_E = k_4/k_3$. This type of model belongs to the class of models that allow the receptor to float in the membrane (Jacobs & Cuatrecasas, 1976).

Binding of hormone

In the absence of enzyme activation, the hormone binding species are R and RE. Thus

$$HRE = HR \cdot E/K_E = H \cdot R \cdot E/k_H K_E, \tag{49}$$

so that

$$HR = H \cdot R/K_H. \tag{50}$$

Consequently the total amount of bound hormone $B \equiv HR + HRE$ is given by

$$B = R[(H/K_H) + (E \cdot H/K_E K_H)]. \tag{51}$$

The conservation equations are

$$E_T = E + HRE, \tag{52}$$

$$R_T = R + HR + HRE. \tag{53}$$

From (53)

$$R = R_T - B. \tag{54}$$

Moreover (52) and (49) yield

$$E_T = E[1 + (HR/K_H K_E)], \tag{55}$$

that is

$$E = E_T K_H K_E/(K_H K_E + H \cdot R). \tag{56}$$

Inserting (51) into (53) and rearranging, one obtains

$$B^2 - B\left[\frac{K_E K_H H + 2H^2 R_T + K_H^2 K_E + (H \cdot R_T + H \cdot E_T K_H)}{H(K_H + H)}\right.$$

$$\left. + \frac{H \cdot R_T K_H K_E + H \cdot E_T \cdot R_T K_H + H^2 (R_T)^2}{H(K_H + H)}\right] = 0. \quad (57)$$

Simulation of (57) reveals negative cooperativity in binding (Figure 7.1). The tighter the interaction between HR and E, the further away are the two apparent binding constants derived for H from the thermodynamic dissociation constant.

The rate of HRE' formation

Using the scheme presented in (48) one can derive the rule of HRE' accumulation (in the presence of Gpp(NH)p) as a function of time. We see that

$$dHRE'/dt = k_5 HRE. \quad (58)$$

Applying the steady state conditions for HRE one obtains

$$dHRE/dt = 0 = k_3 HR \cdot E - (k_4 + k_5)HRE \quad (59)$$

or

$$HRE = k_3 HR \cdot E / (k_4 + k_5). \quad (60)$$

But since

$$E_T = E + HRE + HRE', \quad (61)$$

$$R_T = R + HR + HRE, \quad (62)$$

and since $HRE \ll E + HRE'$, (61) and (62) simplify to

$$E_T = E + HRE', \quad (63)$$

and

$$R_T = R + (H \cdot R/K_H) + HRE'. \quad (64)$$

Therefore

$$R = \frac{(R_T - HRE')K_H}{K_H + H}, \quad (65)$$

$$HR = \frac{H \cdot R}{K_H} = \left\{ \frac{H(R_T - HRE')}{K_H(K_H + H)} \right\} K_H = \frac{H(R_T - HRE')}{H + K_H}, \tag{66}$$

and

$$HRE = \frac{k_3 H}{(k_4 + k_5)(K_H + H)} (R_T - HRE')(E_T - HRE'). \tag{67}$$

Since the rate of accumulation of HRE' is given by (48), one obtains

$$\frac{dHRE'}{HRE'^2 + (R_T + E_T) \cdot HRE' - R_T \cdot E_T} = \frac{k_5 k_3 H}{(k_4 + k_5)(K_H + H)} dt. \tag{68}$$

Integration of (68) can be carried out with the aid of any standard table of integrals. The result is, when $R_T < E_T$,

$$HRE' = \frac{E_T \cdot R_T \{1 - \exp(-[k_3 k_5 H(E_T - R_T)/(k_4 + k_5)(K_H + H)]t)\}}{E_T - R_T \{\exp(-[k_3 k_5 H(E_T - R_T)/(k_4 + k_5)(K_H + H)]t)\}}. \tag{69}$$

When $R_T > E_T$ one should interchange E_T and R_T in (69). From (69) it is apparent that the kinetics of HRE' formation deviate strongly from first-order kinetics. When $k_5 < K_4$ HRE will rapidly accumulate *prior* to activation in an equilibrium fashion. We can therefore use scheme (20) to derive an expression for HRE.

$$HRE = \frac{H \cdot R_T(E_T - HRE' - HRE)}{K_E(K_H + H) + H(E_T - HRE' - HRE)} \tag{70}$$

The full expression for HRE is then

$$HRE = \frac{E_T - HRE' + R_T}{2} + \frac{K_E(K_H + H)}{2H} - \sqrt{x}, \tag{71}$$

where

$$x = \{[(E_T - HRE' + R_T)H + K_E(K_H + H)]^2/2H\} - (E_T - HRE')R_T.$$

We can now proceed to obtain an expression for HRE' accumulation using (58).

$$dHRE'/dt = k_5 HRE. \tag{72}$$

Upon defining

$$C = [K_E(K_H + H) + (E_T + R_T)H]/2H, \tag{73}$$

we obtain

$$\frac{dHRE'}{dt} = k_5\left\{\left(\frac{-HRE'}{2}\right) + C - \sqrt{y}\right\},\tag{74}$$

where

$$y = [(HRE'^2/4) - HRE'(R_T + C) + C^2 - E_T \cdot R_T].$$

It can be seen that the rate law which governs the formation of the active species HRE' deviates strongly from first-order kinetics. A solution identical to (69) is obtained when the activation follows the scheme

$$H + R \underset{k_2}{\overset{k_1}{\rightleftarrows}} HR:$$

$$HR + E \underset{k_4}{\overset{k_3}{\rightleftarrows}} HRE \overset{k_5}{\longrightarrow} HRE' \underset{k_7}{\overset{k_6}{\rightleftarrows}} HR + E'.\tag{75}$$

Equation (69) (where $R_T \neq E_T$) describes a second-order process of accumulation of the activated species HRE'.

In most hormone adenylate cyclase systems $R_T = E_T$ (Levitzki *et al.*, 1975). When $R_T = E_T$ the formulation of the process is different but the nature of the process remains second order. Under these conditions (68) can be written in the form

$$\frac{dHRE'}{HRE'^2 + 2E_T \cdot HRE + (E_T)^2} = \frac{k_5H}{K_E(K_H + H)}\,dt.\tag{76}$$

Integrating (76), we find that

$$\frac{1}{HRE'} = \frac{K_E(K_H + H)}{(E_T)^2 k_5 Ht} + \frac{1}{E_T}$$

or

$$\frac{1}{HRE'} = \frac{K_E(K_H + H)}{(R_T)^2 + k_5 Ht} + \frac{1}{R_T}.\tag{77}$$

Equation (77) reveals that the kinetics of HRE' accumulation are second order. In particular a plot of $1/HRE'$ versus $1/t$ should yield a straight line. It is therefore also clear that plotting the data according to first-order kinetics of activation, namely plotting $\ln\left[(HRE'_{max} - HRE'_t)/HRE'_{max}\right]$ versus t, will yield a nonlinear dependence. From (69) it is apparent that when $E_T \gg R_T$ the rate of appearance of HRE' approaches a first-order process. Similarly when $R_T \gg E_T$ the rate of accumulation of HRE' becomes first order.

In summary, the equilibrium floating model which assumes that the receptor–hormone complex and the enzyme are at equilibrium with the hormone–receptor–enzyme (HRE) species predicts that (*a*) the binding of hormone is negatively cooperative; (b) the overall rate process of appearance of the permanently active state of the enzyme is second order at every hormone concentration; (c) no explicit expression for the dependence of the apparent rate constant, characterizing the conversion of the inactive enzyme to its permanently active form can be formulated; (d) reduction in either R_T or E_T by chemical modification will reduce both the rate at which HRE' is produced and the maximal activation level attainable.

Another variation of the model presented in (75) is the following:

$$H + R \underset{K_H}{\rightleftharpoons} HR; \quad HR + E \rightleftharpoons HRE \rightarrow HRE' \rightleftharpoons HR + E'$$

$$K_H \Updownarrow \tag{78}$$

$$H + RE'$$

This model is very similar to the one analyzed in detail and also predicts negative cooperativity in hormone binding and complex kinetics of accumulation of the active species of the enzyme (HRE', RE', and E'). In fact, the addition of any number of equilibrium steps subsequent to the irreversible step of activation does not fundamentally change the rate of activation. The receptor enzyme equilibrium model analyzed by Jacobs & Cuatrecasas (1976) is similar to the models discussed in this section, but less general. In their analysis, Jacobs & Cuatrecasas assumed that HRE is the active species of the enzyme, which is different from the more general case presented here. Also these authors included another simplifying assumption: that the hormone binds to the naked receptor and to the receptor–enzyme complex with identical affinities. The consequences of the latter assumption are (a) that the binding pattern of the hormone is noncooperative (derivation not shown) and (b) the kinetics of enzyme activation by hormone and Gpp(NH)p deviate from first-order kinetics (derivation not shown).

The collision coupling model (IV)

In this model, the mode of enzyme activation is described by the scheme

$$H + R \underset{K_H}{\rightleftharpoons} HR + E \underset{k_4}{\overset{k_3}{\rightleftharpoons}} X \overset{k_5}{\longrightarrow} HR + E' \tag{79}$$

where $K_H = k_2/k_1$ and X is the ternary encounter complex between hormone, receptor and enzyme. During the lifetime of this complex the enzyme is converted to its activated form E' such that X = HRE $\rightleftharpoons$ HRE'.

The collision coupling model is in fact a special case of the floating equilibrium model, model III. In its formulation a few assumptions are made: (i) k_4 is very arge, so that the affinity of HR for E is small, and thus k_3

becomes rate limiting; (ii) k_5 is also very rapid compared to k_3 and is of the same order of magnitude as k_4, $k_5 > k_4$; (iii) HRE never accumulates. In this case the irreversible step of activation (in the presence of Gpp(NH)p) is no longer between HRE and HRE', but describes the transition between HR and E since $k_5 \geq k_4$ and both are extremely fast. In analogy with the interaction of an enzyme with substrate, E is the substrate and R is the catalytic entity that is regenerated concomitantly with the formation of the 'product' E'.

Hormone binding

The binding pattern of hormone in this case is simple since the only bound hormone species that exists in significant amounts is HR. Therefore

$$HR = R_T H/(K_H + H), \tag{80}$$

and the binding of hormone is noncooperative (Figure 7.1).

The kinetics of appearance of permanently activated state

For the model presented in (79) one can write

$$E_T = E + X + E', \tag{81}$$

$$R_T = R + R + (HRE \rightleftarrows HRE') + HR. \tag{82}$$

Also

$$(HRE + HRE') \ll RE + E + E', \tag{83}$$

and

$$(HRE + HRE') \ll HR + RE + R. \tag{84}$$

One can write

$$dE'/dt = k_5 X, \tag{85}$$

so that

$$dX/dt = k_3 HR \cdot E - (k_4 + k_5)X. \tag{86}$$

At the steady state, which is rapidly established,

$$X = k_3 HR \cdot E/(k_4 + k_5). \tag{87}$$

Since

$$HR = R_T \cdot H/(K_H + H), \tag{88}$$

and

$$E = E_T - E', \tag{89}$$

one can rewrite (85) as

$$\frac{dE'}{dt} = \frac{k_5 k_3 R_T \cdot H \cdot E_T}{(k_4 + k_3)(K_H + H)} - \frac{k_5 k_3 R_T \cdot H \cdot E'}{(k_4 + k_3)(K_H + H)}, \tag{90}$$

so that

$$\frac{dE'}{E_T - E'} = \frac{k_5 k_3 R_T \cdot H}{(k_4 + k_5)(K_H + H)} dt. \tag{91}$$

Integrating (91) one obtains:

$$E' = E_T\{1 - \exp\left(-[k_3 k_5 R_T \cdot H/(k_4 + k_5)(K_H + H)]t\right)\}. \tag{92}$$

In contrast to model III, the rate constant governing the appearance of the activated state depends solely on the total receptor concentration and not on both the total receptor and total enzyme concentrations. The maximal level of activated enzyme attainable is solely dependent on the total enzyme concentration and *independent* of the receptor concentration. The overall process of enzyme activation is first order and the apparent first-order rate constant is linearly dependent on total receptor concentration and increases as a function of hormone concentration in a noncooperative fashion.

Activation of the enzyme in the presence of GTP

In the presence of GTP the activated form of the enzyme is reverted to its inactive form concomitantly with the hydrolysis of GTP at the regulatory site (Cassel & Selinger, 1976; Levitzki, 1977; Sevilla & Levitzki, 1977). Under these conditions the collision coupling model is given by the scheme

$$H + R \underset{K_H}{\rightleftharpoons} HR + E \underset{k_4}{\overset{k_3}{\rightleftharpoons}} (HRE \rightleftharpoons HRE') \overset{k_5}{\longrightarrow} HR + E' \overset{k_6}{\longrightarrow} E. \tag{93}$$

The equation governing the rate of appearance of E' turns out to be

$$E' = \frac{E_T k_5 k_3 R_T \cdot H/(k_4 + k_5)(K_H + H)}{\{k_5 k_3 R_T \cdot H/(k_4 + k_5)(K_H + H)\} + k_6}$$

$$\times \left\{1 - \exp\left(-\left[\frac{k_3 k_5 R_T \cdot H}{(k_4 k_5)(K_H + H)} + k_6\right]t\right)\right\}. \tag{94}$$

At high hormone concentration and if $k_5 \gg k_4$, (94) takes the form

$$E' = \frac{E_T}{1 + (k_6/k_3 R_T)} \{1 - \exp(-(k_3 R_T + k_6)t)\}. \tag{95}$$

After steady state conditions have been achieved, the concentration of enzyme in its active form is given by

$$E' = \frac{E_T}{1 + (k_6/k_3 R_T)}. \tag{96}$$

As in model I, it can be seen that only a fraction of the total enzyme is in its active form when the physiological regulator GTP is present. Also as in model I, when Gpp(NH)p is present as the allosteric activator, all the enzyme is converted to its active form after enough time has elapsed. This model predicts that (*a*) hormone binding is noncooperative; (*b*) the kinetics of activation of the enzyme in the presence of saturating Gpp(NH)p is first order; (*c*) the apparent first-order rate constant describing the process of activation is linearly dependent on total receptor concentration and is independent of enzyme concentration. Thus irreversible inactivation of the receptor will cause a proportional decrease in the rate of enzyme activation but not in the maximal activity attainable in the presence of Gpp(NH)p. Irreversible inactivation of the enzyme will cause a decrease in the maximal activity attainable but not affect the rate constant of activation in the presence of Gpp(NH)p. The model also predicts that (*d*) the apparent first-order rate constant of activation of the enzyme in the presence of Gpp(NH)p increases as a function of hormone concentration in a saturable fashion according to a noncooperative dependence curve.

Combination models (V)

One can combine the models discussed so far and construct combination models. One such model is presented in the following scheme:

$$
\begin{array}{ccc}
\mathrm{H + RE} \underset{K_H}{\rightleftharpoons} \mathrm{X} \xrightarrow{k} \mathrm{HR + E'}, \\[2mm]
K_E \big\Updownarrow \qquad\qquad K_H \big\Updownarrow \\[2mm]
\mathrm{R + E} \qquad\qquad \mathrm{H + R}
\end{array}
\tag{97}
$$

where $\mathrm{X = HRE \rightleftharpoons HRE'}$. This model can be termed the precoupled-dissociation model.

Hormone binding

The conservation equations are

$$R_T = HRE + RE + HR, \tag{98}$$

$$E_T = E + RE + HRE. \tag{99}$$

The total amount of bound receptor is given by

$$B \equiv HR + HRE = R_T - RE - R, \tag{100}$$

i.e.

$$B = \frac{H \cdot R}{K_H} + \frac{H \cdot RE}{K_H} = \frac{H}{K_H}(R_T + RE). \tag{101}$$

From (100) and (101) we obtain

$$B = (H/K_H)(R_T - B). \tag{102}$$

The solution for B is

$$B = R_T H/(K_H + H). \tag{103}$$

Thus the binding of hormone under these conditions is noncooperative (Figure 7.1).

The kinetics of appearance of the activated enzyme

In the presence of Gpp(NH)p the scheme of enzyme activation takes the form

$$
\begin{array}{ccc}
H + RE \underset{k_2}{\overset{k_1}{\rightleftarrows}} X \xrightarrow{k_3} HR + E' \\
K_E \updownarrow \qquad \qquad K_H \updownarrow \\
H + R + E \qquad \quad H + R \\
\updownarrow K_H \\
HR
\end{array}
\tag{104}
$$

where $K_H = k_2/k_1$. We have

$$E_T = E + RE + X + E', \tag{105}$$

$$R_T = R + RE + X + HR, \tag{106}$$

$$dE'/dt = k_3 X. \tag{107}$$

When $k_3 \ll k_2$

$$X \ll E + RE + E', \tag{108}$$

$$X \ll R + HR + RE, \tag{109}$$

and one can write the steady state approximation

$$X = k_1 H \cdot RE/(k_2 + k_3). \tag{110}$$

We wish to derive an expression for RE. To this end we use the relations

$$RE = R \cdot E/K_E, \tag{111}$$

$$R = HRK_H/H, \tag{112}$$

which together with (106) and (110) yield

$$R_T = RE + \frac{REK_{\dot{E}}[(K_H + H)/K_H]}{(E_T - E' - RE)}. \tag{113}$$

Upon rearrangement one obtains

$$RE^2 - RE\left(R_T + E_T - E' + \frac{K_E(K_H + H)}{K_H}\right) + R_T(E_T - E') = 0. \tag{114}$$

An explicit analytical solution can be obtained for conditions where RE is relatively small, namely when

$$RE \ll E' + X + E \tag{115}$$

at all times, so that the quadratic term can be neglected. This will be true when K_E is large relative to R and E and H is abundant. The approximate solution is

$$RE \approx \frac{R_T(E_T - E')}{R_T + E_T - E' + [K_E(K_H + H)/K_H]}, \tag{116}$$

since RE^2 is very small. Using (107) and (110) one obtains

$$\left\{\frac{R_T + E_T + [K_E(K_H + H)/K_H] - E'}{E_T - E'}\right\}\frac{dE'}{dt} = \frac{k_3 k_1}{(k_2 + k_3)} H \cdot R_T.$$

(117)

Integrating (117) and setting $E' = 0$ when $t = 0$, we find that

$$\ln\left[\frac{(E_T - E')}{E_T}\right] = \frac{E'/R_T}{1 + [K_E(K_H + H)/R_T K_H]}$$

$$- \frac{k_3 k_1 H \cdot R_T t}{(k_2 + k_3)[R_T + (K_E/K_H)(K_H + H)]}.$$

(118)

The kinetics of appearance of E′ deviate from a first-order process.

Since (118) was formulated for conditions where K_E is large relative to R_T and E_T we simulated the effect of the term

$$\frac{E'/R_T}{1 + [K_E(K_H + H)/R_T K_H]},$$

(119)

which appears in (118), under conditions where $K_E = R_T$. At low hormone concentrations the contribution of the term (119) is very large at low E' and becomes progressively smaller as E' increases. At high hormone concentrations the contribution of (119) will be negligible at all E' levels. Thus deviations from first-order kinetics will be apparent at low hormone levels, while at saturating hormone levels first-order kinetics will be apparent. As in the collision model (model IV) the rate of E′ appearance is governed by R_T and not by E_T, whereas the maximum level of activity attained is determined solely by E_T.

Experimental results

The turkey erythrocyte adenylate cyclase proved to be a very useful experimental system to investigate the possible modes of receptor to enzyme interactions (Tolkovsky & Levitzki, 1978a, b, c). The enzyme in this membrane system can be activated by either the β-adrenergic receptor upon binding of an *l*-catecholamine (Sutherland *et al.*, 1965), or by the adenosine receptor upon binding of adenosine (Tolkovsky & Levitzki, 1978c). In each of these cases we have followed (*a*) the binding of the hormone to the receptor; (*b*) the kinetics of adenylate cyclase activation by the hormone as a function of hormone concentration; (*c*) the change in the kinetic pattern of cyclase activation as a function of receptor and enzyme concentration.

These three types of experiments were shown above to be useful in deciphering the mode of coupling between the receptor and the enzyme.

Binding experiments

The binding of β-antagonists such as [^{3}H]propranolol and [^{125}I]-hydroxybenzylpindolol was shown to be noncooperative (Levitzki *et al.*, 1974; Atlas, Steer & Levitzki, 1974; Levitzki *et al.*, 1975; Tolkovsky & Levitzki, 1978*a*). Similarly the binding of agonists such as *l*-epinephrine, *l*-norepinephrine and *l*-isoproterenol was also found to be noncooperative. Thus, one can reject the dissociation model (model II, Tables 7.1 and 7.2) and the equilibrium floating model (model III, Tables 7.1 and 7.2) for receptor to cyclase coupling. Models II and III predict negatively cooperative binding (Figure 7.1).

Kinetics of adenylate cyclase activation

The kinetics of accumulation of the active species of adenylate cyclase was found to be pseudo first-order for both the *l*-catecholamine (Tolkovsky & Levitzki, 1978*a*, *b*) and the adenosine induced activation (Tolkovsky & Levitzki 1978*c*; Braun & Levitzki, 1979*a*, *b*). Furthermore, the pseudo first-order rate constant was found to depend on agonist concentration in a Michaelian fashion [see, for example, (12) and (92)]. These findings again exclude the dissociation model (model II, Tables 7.1 and 7.2) and the floating model (model III, Tables 7.1 and 7.2). On the basis of the finding that the kinetics of adenylate cyclase activation by either adenosine or *l*-catecholamines is first order even at very low hormone concentrations, the precoupled-dissociation model (model V, Tables 7.1 and 7.2) can also be rejected. On the same grounds the model proposed by Jacobs & Cuatrecasas (1976), which predicts non-first-order kinetics of activation, can also be rejected.

The kinetic pattern as a function of receptor concentration

Our findings so far can be accounted for by two diametrically opposed models: the precoupled model (model I, Tables 7.1 and 7.2) and

Table 7.1. *Mode of hormone binding as predicted by different models or receptor to enzyme coupling*

Model	Formulation	Mode of binding	Percent binding in each class of sites	Ratio of apparent hormone dissociation constant to K_H: K_{app}/K_H
I. Precoupled	$RE + H \underset{K_H}{\rightleftharpoons} HRE$	Noncooperative		1.0
II. Dissociation	$RE + H \underset{K_H}{\rightleftharpoons} HRE \underset{K_E}{\rightleftharpoons} RH + E$	Negatively cooperative. Cooperativity increases as K_E decreases relative to R_T or ET_T	1. 27% of one class and 73% of the second class when $K_E = E_T = R_T$	0.017 0.30
			2. 9% of one class and 81% of another class when $K_E = 0.1E_T = 0.1R_T$	0.013 0.64
III. Equilibrium floating	$H + R \underset{K_H}{\rightleftharpoons} HR + E \underset{K_E}{\rightleftharpoons} HRE$	Negatively cooperative	1. 70% of one class and 30% of another class when $K_E = E_T = R_T$	0.33 0.31
			2. 65% of one class and 35% of another class when $K_E = 0.1E_T = 0.1R_T$	0.062 0.25
IV. Collision coupling	$H + R \underset{K_H}{\rightleftharpoons} HR + E \rightleftharpoons HRE$	Noncooperative	100	1.0
V. Precoupled-dissociation	$H + RE \underset{K_H}{\rightleftharpoons} HRE$ $K_E \updownarrow \qquad \updownarrow K_H$ $H + R + E \underset{K_H}{\rightleftharpoons} HR$	Noncooperative	100	1.0

Table 7.2. *Dependence of the rate of enzyme activation on hormone concentration as predicted by the different models of receptor to enzyme coupling*

| Model | Overall kinetic features | | | Effect of reducing E_T | Effect of reducing R_T |
	Rapid equilibrium prior to activation	Rapid activation compared to equilibrium	Type of dependence of apparent first-order rate constant on hormone concentration		
I. Precoupled	First order	First order	Noncooperative	1. Reduction of maximal activity attainable but no change in rate of activation	1. Reduction in maximal binding and in maximal activity attainable 2. No change in rate of activation
II. Dissociation	Deviates from first order	Deviates from first order	—*	1. Reduction of maximal activity attainable and in activation rate	1. Reduction in maximal binding and in maximal activity attainable
III. Equilibrium floating	Deviates from first order	Second order	—*	1. Reduction as in model II.	1. As in model II
IV. Collision coupling	Model excludes this condition	First order	Noncooperative	1. Reduction in maximal extent of enzyme activation with no change in rate constant for activation	1. Reduction in rate constant of enzyme activation with no change in maximal extent of enzyme activation
V. Precoupled-dissociation	Not examined	Deviates from first order at low hormone concentrations	—*	1. Reduction in extent of enzyme activation	1. As in model IV

* Due to deviations from first-order kinetics no dependence of the rate constant of enzyme activation on hormone concentration will be meaningful.

the collision coupling model (model IV, Tables 7.1 and 7.2). Both models predict noncooperative binding of first-order kinetics of enzyme activation by the agonist. By further analysis of the two models, we have already established the different predictions of the two models as well as the similarities between the two (see also Tables 7.1 and 7.2). We can summarize the predictions of the two models as follows:

The precoupled model (model I) predicts that the apparent rate constant, k_{obs}, is independent of the total receptor concentration, whereas the final concentration of attainable activated enzyme is proportional to the latter (12). The collision coupling model (model IV) in contrast, predicts that the apparent rate constant k_{obs}, is linearly dependent upon the total receptor concentration, R_T, whereas the final concentration of attainable activated enzyme is independent of this quantity (92).

A diagnostic experiment was therefore designed in order to distinguish between the two possibilities.

The receptor concentration was progressively reduced by reacting the membrane preparation with an irreversible blocker which reacts covalently with the receptor. The progressive decrease in the concentration of the β-adrenergic receptor using a specific β-receptor directed affinity label resulted in a proportional decrease in the pseudo first-order rate constant whereas the maximal level of adenylate cyclase activation remains unaffected (Tolkovsky & Levitzki, 1978a, b). A summary of these experiments is shown in Figure 7.2. It is clear therefore that the mode of coupling of adenylate cyclase to the β-adrenergic receptor is of the 'collision coupling' type (model IV, Tables 7.1, 7.2).

In contrast, a similar experiment using an adenosine affinity label revealed that the adenosine receptor is tightly coupled to the adenylate cyclase (Braun & Levitzki, 1979a, b). The progressive destruction of the adenosine receptor by the adenosine affinity label results in a proportional decrease in the maximal level of enzyme activity attainable whereas the pseudo first-order rate constant of activation remains unchanged (Braun & Levitzki, 1979a, b). Other kinetic experiments corroborate the latter finding. For example, the pattern of adenylate cyclase activation by the combined ligands adenosine and epinephrine has been analyzed (Tolkovsky & Levitzki, 1978c). Knowing that the mode of coupling between the β-receptor and the cyclase is by the 'collision coupling' mechanism one can deduce the mode of coupling of the adenosine receptor to the enzyme from the kinetics of enzyme activation by the two ligands combined. This analysis (Tolkovsky & Levitzki, 1978c) revealed that the adenosine receptor is permanently coupled to the enzyme. In summary, the mode of coupling of adenosine to adenylate cyclase corresponds to the precoupled model.

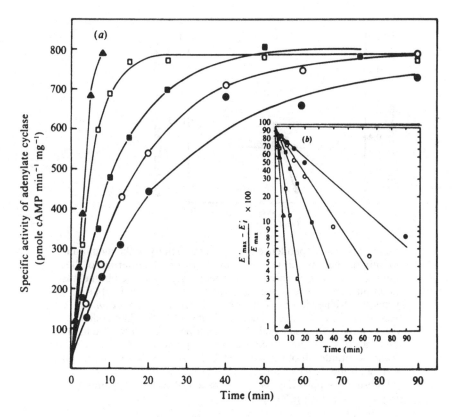

Figure 7.2. Time course of adenylate cyclase activation by saturating Gpp(NH)p and *l*-epinephrine subsequent to affinity labeling treatment. The membranes were treated with the affinity label as described elsewhere (Tolkovsky & Levitzki, 1978*b*). The rate of enzyme activation to its permanently active state was measured in the presence of 1.0×10^{-4} mole l^{-1} *l*-epinephrine and $1.0 \times 10^{-4} l^{-1}$ Gpp(NH)p (saturating concentrations). (*a*) Progress curves for the accumulation of the active enzyme. E'_{max} is the maximal specific activity attained which was equal in this particular preparation to 790 pmoles cAMP mg^{-1} min^{-1}. (*b*) Data in (*a*) plotted on a semilogarithmic plot to demonstrate that the process of enzyme activation is first order. E'_{max} is the maximal specific activity attained, and E'_i the specific activity in each of the time points. ▲, untreated membranes; □, 1.67×10^{-5} mole l^{-1} affinity label; ■, 4.3×10^{-5} mole l^{-1} affinity label; ○, 1.0×10^{-4} mole l^{-1} affinity label; ●, 2.3×10^{-4} mole l^{-1} affinity label.

Other experimental approaches

Since the activation of adenylate cyclase by the β-receptor is a bimolecular process, one expects that the kinetics of enzyme activation by the hormone-bound reeptor should depend strongly on the viscosity of the

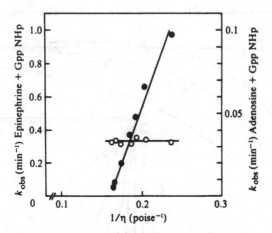

Figure 7.3. The dependence of k_{obs} on the fluidity of the membrane. Membrane fluidity was increased by the controlled insertion of the fluidizing agent *cis*-vaccenic acid. The fluidity of the membrane was determined by fluorescence polarization using 1,6-diphenyl-(1,3,5 all trans)-hextriene (DPH) as described elsewhere (Hanski *et al.*, 1978; Rimon *et al.*, 1978). In parallel, the kinetics of adenylate cyclase activation by adenosine and by epinephrine were determined in the presence of saturating concentration of Gpp(NH)p. k_{obs} was calculated from a computer fit of the data to (12) for adenosine and to (92) for *l*-epinephrine. ○, rate constant of cyclase activation by adenosine + Gpp(NH)p. ●, rate constant of cyclase activation by *l*-epinephrine + Gpp(NH)p.

membrane matrix. Indeed, it has recently been shown (Hanski, Rimon & Levitzki, 1979) that the rate constant of enzyme activation by the hormone bound β-receptor depends linearly on the fluidity of the membrane. It appears that the bimolecular reaction between the receptor and the enzyme is diffusion controlled. As expected, on the basis of the experiments described in the previous section, the rate of adenylate cyclase activation by the adenosine receptor is independent of membrane fluidity. A summary of these results is given in Figure 7.3. In fact, one can elucidate the mode of adenylate cyclase activation by hormone receptors by examining the kinetics of enzyme activation by the receptor as a function of membrane fluidity. The fluidity of the membrane can be increased by inserting fatty acids (Hanski *et al.*, 1979) and phospholipids (Hanski & Levitzki, 1978) or decreased by the insertion of cholesterol (Klein, More & Pastan, 1978).

Discussion

The mechanisms of receptor to adenylate cyclase coupling were explored in detail. Different modes of enzyme to receptor coupling within the membrane predict both different patterns of hormone binding and different mechanisms of activation of the receptor dependent enzyme. A summary of the important predictions of the different models is given in Tables 7.1 and 7.2 and in Figure 7.1. The possibility of generating conditions under which the process of hormone-dependent enzyme activation is essentially irreversible simplifies the mathematical analysis and enhances the possibility of designing experiments to probe the mode of receptor to enzyme coupling. Thus, performing binding experiments in conjunction with detailed kinetic analysis of enzyme activation may allow one to decide which of the possible modes of coupling operates *in vivo*.

Since numerous hormone-dependent adenylate cyclases have been shown to depend also on GTP and to respond to GPP(NH)p, the analysis can be extended to other hormone-dependent adenylate cyclases.

In some of the models explored, the response in the presence of GTP, instead of Gpp(NH)p, was also analyzed. Under such conditions not all the enzyme is converted to its activated form but a steady state concentration of active enzyme is achieved. This latter situation is the one encountered under physiological conditions. The predictions of the different models as to the behavior of the system under such conditions can also be tested against experimental findings.

The formulation of the coupling between enzyme and receptors within a membrane matrix is general and may apply to other receptor dependent responses where both response and ligand binding to the receptor may be studied simultaneously.

Progress in the 1980s: the role of adenylate cyclase sub-units in the activation and inhibition of the enzyme

During the 1980s it was firmly established that the catalytic unit C of the enzyme is activated by the stimulatory GTP binding protein G_s and is inhibited by the inhibitory GTP binding protein G_i. G_s is composed of three subunits, α which harbors the GTP binding site, β and γ. G_i is also composed of three subunits α_i which harbor the GTP binding site, β and γ. The subunits α_s and α_i are homologous and β and γ are identical in G_s and G_i (see Gilman, 1987 and Levitzki, 1986 for reviews).

Stimulatory receptors like the β-adrenergic receptor interact with G_s, stimulating cyclase, and inhibitory hormones interact with inhibitory receptors that interact with G_i, inhibiting cyclase. Upon interaction of a stimulatory hormone receptor with G_s, in its hormone bound form (HR), the α_s-subunit 'opens' up (Braun, Tolkovsky & Levitzki, 1982), GDP falls out, and GTP binds. As a result of this binding α_s and therefore G_s is activated, in turn activating the catalyst C. GTP is then hydrolyzed by GDP and P_i, leading to enzyme deactivation. The cycle repeats itself as long as hormone and GTP are present. This description tells us that the entity E described above is actually a complex between G_s and C.

Shuttle models

The biochemical observation that E is composed of G_s and C immediately suggested to some researchers (Citri & Schramm, 1980; DeLean, Stadel & Lefkowitz, 1980) that G_s and C are separate units and the sequence of events is as follows:

$$HR + G_s(GDP) \rightleftharpoons HRG_s'(GDP)$$

$$HRG_s'(GDP) + GTP \rightleftharpoons HRG_s''(GTP) + GDP$$

$$HRG_s''(GTP) \rightleftharpoons HR + G_s''(GTP)$$

$$G_s''(GTP) + C' \rightleftharpoons G_s''(GTP)C$$

$$H_2O + G_s''(GTP)C' \longrightarrow G_s'''(GDP)C' + P_i$$

$$G_s'''(GDP)C' \rightleftharpoons G_s(GDP) + C.$$

This 'shuttle' type model still preserves the catalytic role of the receptor characteristic for the 'collision coupling' (model IV above). However, the kinetics of cyclase activation is not linear with receptor concentration and the rate depends on the concentration of G_s and C. Other more elaborate 'shuttle' models also predict complex kinetics of cyclase activation. It was therefore concluded that the experimentally observed results that led to the suggestion of 'collision coupling' originally, cannot be accommodated with such models (Tolkovsky & Levitzki, 1981). It was necessary to infer that G_s and C must be permanently associated. Kinetic experiments similar to those described above were immediately performed, but this time varying the concentration of G_s and C within the native membrane. Reducing the concentration of C by chemical inactivation, and G_s by limiting the amount of GPPNHP such that the product $[G_s][C]$ was reduced down to the experimentally accessible limit (200 fold), did not change the basic kinetic features of the system. (i) The overall kinetics of activation by hormone and GPPNHP is first order. (ii) This kinetic feature is preserved at all receptor occupancies tested.

These results confirmed the previous hypothesis that the enzyme depicted originally as E is G_sC where G_s and C remain permanently associated both as $G_s(GDP)C$ and $G_s(GTP)C$ (Tolkovsky, Braun & Levitzki, 1982). It was later shown that G_sC can be purified as such both at the GDP bound state and the *GPPNHP* bound state (Arad, Rosenbusch & Levitzki, 1984), thus confirming directly the kinetic results of Tolkovsky *et al.* (1982). More recently, it has been shown in a system reconstituted from pure β-adrenergic receptor, pure G_s, and pure C that the overall kinetic features are those of 'collision coupling' as originally suggested (Hekman *et al.*, 1984; Feder *et al.*, 1986*a*, *b*). Other groups have recently confirmed these findings (summarized in Levitzki, 1988).

The issue of G-subunit dissociation

The non-hydrolyzable *GTP* analog *GTP S* and high Mg^{2+}, in the presence of the detergent Lurbol-PX, induce the irreversible dissociation of G_s:

$$G_s(GDP) + GTP\gamma S \rightarrow G_s(GTP\ S) + \beta\gamma$$

This finding suggested to Gilman (1987) that hormonal stimulation in the presence of the natural ligand *GTP* leads to the reversible dissociation of G_s to $\alpha_s(GTP)$, which then becomes associated with the catalytic unit C, activating the latter, and $\beta\gamma$. Upon hydrolysis of *GTP* to *GDP*, $\alpha_s(GDP)$ dissociates from C and reassociates with the $\beta\gamma$-subunits. The $\beta\gamma$-subunits, according to this scheme, compete with C for α_s. Hence, elevation of the level of the $\beta\gamma$-subunits caused by activation of G_i by inhibitory receptors scavenges α_s, inhibiting its interaction with C.

This feature of the model ties in cyclase inhibition and assigns a regulatory role to $\beta\gamma$. In itself, this is an elegant model that may explain both enzyme activation by G_s and enzyme inhibition by G_i, lending $\beta\gamma$ a key regulatory role. This model also accounts for the failure to demonstrate a G_i-to-C interaction in reconstituted systems, and provides a mechanism by which G_i confers inhibition of C without direct physical interaction. However, it was suggested, and in fact evidence for this idea was provided, that G_i probably interacts physically with C, but only in the presence of G_s (Marbach, Shiloach & Levitzki, 1988).

The scheme suggesting that α_s dissociates from G_s and *then* couples to C also actually suggests that G_s is a separate entity from C in the membrane bilayer and implies that the α_s subunit shuttles between G_s and the catalytic unit C. It has, however, already been demonstrated that any mechanism that implies G_sC dissociation cannot be compatible with the kinetic data in native membranes, as well as in reconstituted systems. A modified G_s dissociation model was therefore suggested (Levitzki, 1984, 1986, 1987), which can be accommodated with the 'collision coupling' model. The modified G_s dissociation scheme suggests that both species, $G_s(GTP)C$ and $\alpha_s(GTP)C$, represent active cyclase molecules, and the $G_s(GTP)C$ species is the less active species. That is, allowing $\beta\gamma^s$ dissociation from the $G_s(GTP)C$ complex does not contradict the experimental findings and in fact preserves the essential features of the 'collision coupling' mechanism.

Current experiments are concerned with how tight the $\beta\gamma$ to α_s association is and whether *indeed* β dissociates upon adenylyl cyclase activation by the hormone bound β-adrenergic receptor, under physiological conditions. In the case of bovine brain adenylate we have recently demonstrated that $\beta\gamma$ subunits remain associated with the GPPNHP activated enzyme (Marbach *et al.*, 1990). Similar results have more recently been obtained for the turkey erythrocyte adenylate cyclase (Bor-Sinai *et al.*, 1991). These do support the assertion that $\beta\gamma$ subunits probably do not mediate cyclase inhibition. These experiments are of importance since it is not at all clear whether β mediates adenylyl cyclase inhibition alone or α_i also participates in the inhibition action.

References

Arad, H., Rosenbusch, J. & Levitzki, A. (1984). The stimulatory GTP regulatory unit Ns and the catalytic unit of adenylate cyclase are tightly associated, mechanistic consequences. *Proc. Nat. Acad. Sci.*, *USA* **81**, 6579–83.

Atlas, D., Steer, M. L. & Levitzki, A. (1971). Stereospecific binding of propanolol and catecholamines to the β-adrenergic receptor. *Proc. Nat. Acad. Sci.*, *USA* **71**, 4246–8.

Bor-Sinai, A., Marbuch, I., Shorr, R. G. L. & Levitzki, A. (1991). The Gβ subunit is permanently associated with turkey erythrocyte adenylyl cyclase. *J. Biol. Chem.* (submitted).

Boeynams, J. M. & Dumont, J. E. (1975). Quantitative analysis of the binding of ligands to their receptors. *J. Cyclic Nucleotide Res.* **1**, 123–42.

Braun, S. & Levitzki, A. (1979a). The attenuation of epinepherine dependent adenylate cyclase by adenosine and the characteristics of the adenosine stimulatory and inhibitory sites. *Mol. Pharmacol.* **16**, 737–48.

Braun, S. & Levitzki, A. (1979b). Adenosine receptor permanently coupled to turkey erythrocyte adenylate cyclase. *Biochemistry* **18**, 2134–8.

Braun, S., Tolkovsky, A. M. & Levitzki, A. (1982). Mechanisms of control of the turkey erythrocyte β-adrenoceptor dependent adenylate cyclase by guanyl nucleotide: A minimum model. *J. Cyclic Nucleotide Res.* **7**, 139–50.

Cassel, D. & Selinger, Z. (1976). Catecholamine stimulated GTPase activity in turkey erythrocyte membranes. *Biochim. Biophys. Acta* **452**, 538–51.

Citri, Y. & Schramm, M. (1980). Resolution, reconstitution and kinetics of the primary action of a hormone receptor. *Nature* **287**, 297–300.

DeLean, A., Stadel, J. J. & Lefkowitz, R. J. (1980). A ternary complex model explains the agonist specific binding properties of adenylate cyclase couples β adrenergic receptor. *J. Biol. Chem.* **255**, 7108–17.

Feder, D., Im, M. J., Pfeuffer, T., Hekman, M., Helmreich, E. J. M. & Levitzki, A. (1986a). The hormonal regulation of adenylate cyclase. *J. Biochem. Soc. Symp.* **52**, 145–51.

Feder, D., Im, M.-J., Hekman, M., Klein, H., Holzhofer, C., Helmreich, E. J. M., Levitzki, A. & Pfeuffer, T. (1986b). Reconstruction of β1-adrenoceptor-dependent adenylate cyclase from purified components. *EMBO J.* **5**, 1509–14.

Gilman, A. G. (1987). The molecular biology of receptor generated signals. *Ann. Rev. of Biochemistry* **56**, 615.

Hanski, E. & Levitzki, A. (1978). The absence of desensitization in the beta adrenergic receptors of turkey reticulocytes and erythrocytes and its possible origin. *Life Sci.* **22**, 53–60.

Hanski, E., Rimon, G. & Levitzki, A. (1979). Adenylate cyclase activation by the β-adrenergic receptors as a diffusion-control process. *Biochemistry* **18**, 846–53.

Hekman, M., Feder, D., Keenan, A. K., Gal, A., Klein, H. U., Pfeuffer, T., Levitzki, A. & Helmreich, E. J. M. (1984). The reconstitution of the β-receptor dependent adenylate cyclase from its purified components. *EMBO J.* **3**(13), 3339–45.

Jacobs, S. & Cuatrecasas, P. (1976). The mobile receptor hypothesis and 'cooperativity' of hormone binding. Application to insulin. *Biochim. Biophys. Acta* **433**, 482–95.

Klein, I., More, L. & Pastan, I. (1978). Effect of liposomes containing cholesterol on adenylate cyclase activity of cultured mammalian fibroblasts. *Biochim. Biophys. Acta* **506**, 42–53.

Levitzki, A. (1977). The role of GTP in the activation of adenylate cyclase. *Biochem. Biophys. Res. Comm.* **74**, 1154–9.

Levitzki, A. (1984). Receptor to effector coupling in the receptor-dependent adenylate cyclase system. *J. Rec. Res.* **4**, 339–409.

Levitzki, A. (1986). β-Adrenergic receptors and their mode of coupling to adenylate cyclase. *Physiol. Revs.* **66**, 819–42.

Levitzki, A. (1987). Regulation of adenylate cyclase by hormones and G-proteins. *FEBS Lett.* **211**, 113–18.

Levitzki, A. (1988). From epinephrine to cAMP. *Science* **241**, 800–6.

Levitzki, A., Sevilla, N., Atlas, D. & Steer, M. L. (1975). Ligand specificity and characteristic of the β-adrenergic receptor in turkey erythrocyte plasma membranes. *J. Mol. Biol.* **97**, 35–53.

Levitzki, A., Sevilla, N. & Steer, M. L. (1976). The regulatory control of β-receptor dependent adenylate cyclase. *J. Supramol. Struct.* **4**, 405–18.

Levitzki, A., Steer, M. L. & Atlas, D. (1974). The binding characteristics and number of β-adrenergic receptors on the turkey erythrocyte. *Proc. Nat. Acad. Sci., USA* **71**, 2773–7.

Marbach, I., Shiloach, J. & Levitzki, A. (1988). Gi affects the agonist binding properties of β-adrenoceptors in the presence of Gs. *Eur. J. Biochem.* **172**, 239–46.

Marbach, I., Bor-Sinai, A., Minich, M. & Levitzki, A. (1990). β-subunits copurifies with GPPNHP activated adenylyl cyclase. *J. Biol. Chem.* **265**, 9999–10004.

Pfeuffer, T. & Helmreich, E. J. M. (1975). Activation of pigeon erythrocyte membrane adenylate cyclase by guanyl nucleotide analogues and separation of a nucleotide binding protein. *J. Biol. Chem.* **250**, 867–76.

Rimon, G., Hanski, E., Braun, S. & Levitzki, A. (1978). Mode of coupling between hormone receptor and adenylate cyclase elucidated by modulation of membrane fluidity. *Nature Lond.* **276**, 394–6.

Schramm, M. & Rodbell, M. (1975). A persistent active state of the adenylate cyclase system produced by the combined actions of isoproterenol and guanylyl imidophosphate in frog erythrocyte membranes. *J. Biol. Chem.* **250**, 2232–7.

Sevilla, N. & Levitzki, A. (1977). The activation of adenylate cyclase by *l*-epinephrine and guanylylimidophosphate and its reversal by *l*-epinephrine and GTP. *FEBS Lett.* **76**, 129–34.

Sevilla, N., Steer, M. L. & Levitzki, A. (1976). Synergistic activation of adenylate cyclase by guanylyl imidophosphate and epinephrine. *Biochemistry* **15**, 3493–9.

Spiegel, A. M., Brown, E. M., Fedak, S. A., Woodend, C. J. & Aurbach, G. D. (1976). Holocatalytic state of adenylate cyclase in turkey erythrocyte membranes: formation with guanylylimidophosphate plus isoproterenol without effect on affinity of β receptor. *J. Cyclic Nucleotide Res.* **2**, 47–56.

Sutherland, E. W., Øye, I. & Butcher, R. W. (1965). The action of epinephrine and the role of adenylate cyclase system in hormone action. *Rec. Prog. Horm. Res.* **21**, 623–46.

Swillens, S. & Dumont, J. E. (1976). The mobile receptor hypothesis in hormone action: A general model accounting for desensitization. *J. Cyclic Nucleotide Res.* **3**, 1–10.

Tolkovsky, A. M., Braun, S. & Levitzki, A. (1982). Kinetics of interaction between the β-adrenoceptor, the GTP regulatory protein and the catalytic unit of turkey erythrocyte adenylate cyclase. *Proc. Nat. Acad. Sci.* **79**, 213–17.

Tolkovsky, A. M. & Levitzki, A. (1978a). Collision coupling of the β-adrenergic receptor with adenylate cyclase. In *Hormones and Cell Regulation*, vol. 2, ed. J. Dumont and J. Nunez, Amsterdam, Elsevier/North Holland, pp. 89–105.

Tolkovsky, A. M. & Levitzki, A. (1978b). Mode of coupling between the β-adrenergic receptor and adenylate cyclase in turkey erythrocytes. *Biochemistry* **17**, 3795–810.

Tolkovsky, A. M. & Levitzki, A. (1978c). Coupling of a single adenylate cyclase to two receptors: adenosine and catecholamine. *Biochemistry* **17**, 3811–17.

Tolkovsky, A. M. & Levitzki, A. (1981). Hypothesis: theories and predictions of models describing sequential interactions between the receptor, the GTP regulatory unit, and the catalytic unit of hormone dependent adenylate cyclases. *J. Cyclic Nucleotide Res.* **7**, 139–50.

8

Models for oscillations and excitability in biochemical systems

Introduction

In response to a change in external conditions, such as substrate input or effector levels, open biochemical systems commonly reach a stationary state. Monotone evolution to a stable time-independent regime is, however, by no means the rule. Even in spatially homogeneous conditions, such diverse phenomena as oscillations, excitability, or transitions between multiple stationary states are possible in metabolic pathways. Space-dependent structures such as chemical waves can be obtained when diffusion is coupled to chemical reaction. Nonequilibrium thermodynamics shows that these phenomena constitute various modes of self-organization in systems close to or beyond instability of a stationary state (Nicolis & Prigogine, 1977). The conditions for the occurrence of such *dissipative structures* (Prigogine, 1969) are (i) a critical distance from equilibrium, (ii) a flow of matter or energy through the system, and (iii) appropriate nonlinear kinetic laws. The three conditions for self-organization in nonequilibrium systems are often satisfied in biology. Oscillatory processes indeed occur at all levels where regulation is exerted. This ranges from periodicities in predator–prey populations (May, 1972) to oscillations in metabolic pathways and in the nervous system (Hess & Boiteux, 1971; Chance, Pye, Ghosh & Hess, 1973; Goldbeter & Caplan, 1976; Berridge & Rapp, 1979; Winfree, 1980; Glass & Mackey, 1988).

Excitability is the capability of a system, initially at a stable steady state, to amplify suprathreshold perturbations in a pulsatory manner. Experiments show that excitable chemical systems can give rise, in slightly different conditions, to sustained autonomous oscillations (Winfree, 1972; De Kepper, 1976). The link between the two phenomena has been studied theoretically in models for the nerve membrane (Fitzhugh, 1961), for an enzyme with autocatalytic pH-dependent kinetics (Hahn, Nitzan, Ortoleva & Ross, 1974), and for the Belousov–Zhabotinsky reaction which is the best-known example of a chemical clock (Tyson, 1977). The purpose of this chapter is to analyze the generation of excitable and oscillatory behavior in biochemical systems. A great deal of theoretical work has been devoted to oscillatory enzyme reactions (see the reviews by Goldbeter & Caplan, 1976, and by Tyson & Othmer, 1978). In the following, we shall restrict outselves to the analysis of two biochemical systems for which experimental evidence exists for excitable and/or oscillatory behavior. This will allow comparison of the predictions of theoretical models with the available experimental data.

We shall consider in turn a soluble and a membrane-bound enzymatic system. The first is that of glycolytic oscillations. These periodicities, observed in intact yeast cells as well as in yeast and muscle extracts, are the prototype of metabolic oscillations (Hess & Boiteux, 1971; Goldbeter & Caplan, 1976; Berridge & Rapp, 1979). They have been studied extensively and their mechanism is relatively simple: they result from the positive regulation exerted on phosphofructokinase by a product of the enzyme, ADP, or by AMP which is linked to ADP through the adenylate kinase reaction. The model studied for the phenomenon is that of an allosteric enzyme regulated by positive feedback. The analysis of this model accounts for many experiments on glycolytic oscillations and predicts conditions in which the pulsatory amplification of suprathreshold ADP pulses should be observed in yeast extracts.

In the second part of this chapter, we shall analyze the cyclic AMP (cAMP) signaling system which controls periodic aggregation in the cellular slime mould *Dictyostelium discoideum*. As demonstrated in cell suspension experiments, this membrane-bound signaling system is capable of amplifying extracellular cAMP pulses and of generating autonomously periodic cAMP signals. Here also the mechanism appears to be based on the positive control exerted on adenylate cyclase by the product of the enzyme, cAMP. The study of the phosphofructokinase and adenylate cyclase models throws light on the closely related conditions in which both excitable and oscillatory behavior arise in regulated biochemical systems.

Excitability and oscillations in glycolysis

Experimental observations
Glycolytic oscillations were first discovered in 1964 and have since been extensively reviewed (Pye, 1969; Hess & Boiteux, 1971; Chance *et al.*, 1973; Goldbeter & Caplan, 1976; Berridge & Rapp, 1979). Here we shall only recall the salient features of these periodicities. The most convenient system in which to study the oscillations is that of cell extracts of yeast or muscle. In yeast extracts, which are most commonly utilized, the period of

the phenomenon is of the order of five minutes (Pye, 1969; Hess & Boiteux, 1971); in muscle extracts, the period approaches twenty minutes (Frenkel, 1968; Tornheim & Lowenstein, 1974). When yeast extracts are subjected to a constant input of glycolytic substrate, oscillations are observed for substrate injection rates lying between two critical values (Hess *et al.*, 1969). Similar experiments have been performed in suspensions of intact yeast (Von Klitzing & Betz, 1970). All glycolytic intermediates oscillate with a unique frequency but with various phases; the periodicities can be conveniently observed by recording the fluorescence of NADH.

That the oscillations are produced by phosphofructokinase (PFK) has been demonstrated in numerous ways. An obvious proof is the suppression of periodicity when the PFK step is bypassed in glycolyzing yeast extracts by the injection of fructose-1,6-bisphosphate (FDP) which is the product of PFK (Hess & Boiteux, 1968). Moreover, positive as well as negative effectors of PFK are potent inhibitors of glycolytic oscillations (Frenkel, 1968; Hess, Boiteux & Krüger, 1969). The reconstitution *in vitro* of an oscillating system comprising the complete sequence of glycolytic enzymes from hexokinase to alcohol dehydrogenase confirms that the periodicities originate from this metabolic pathway (Hess & Boiteux, 1968). An oscillating system consisting of a shorter sequence of reactions centered around the PFK step has been described (Hofmann, 1978).

As shown by phase-shift experiments, it is the couple ATP/ADP which controls the periodic change in PFK activity (Hess & Boiteux, 1968; Pye, 1969). The second substrate, fructose-6-phosphate, does not play any significant regulatory role; FDP also activates the enzyme in muscle extracts (Tornheim & Lowenstein, 1974). Models based on the activation of PFK by ADP or FDP have been proposed by Higgins (1964) and Sel'kov (1968); these models account qualitatively for several features of oscillating glycolysis. A model for the oscillatory PFK reaction explicitly taking into account the allosteric properties of the enzyme was subsequently developed (Goldbeter & Lefever, 1972; Goldbeter & Nicolis, 1976). This model yields insight into the molecular mechanism of glycolytic oscillations and allows prediction of the conditions in which excitable behavior should be observed in glycolyzing yeast or muscle extracts. More complex models including additional enzymatic steps have been considered by Termonia & Ross (1981) and Markus & Hess (1984).

Allosteric model for the oscillatory phosphofructokinase reaction

The allosteric model for PFK is based on the concerted transition mechanism proposed by Monod, Wyman & Changeux (1965) for multisubunit enzymes (see Chapter 3). The results obtained in a concerted model

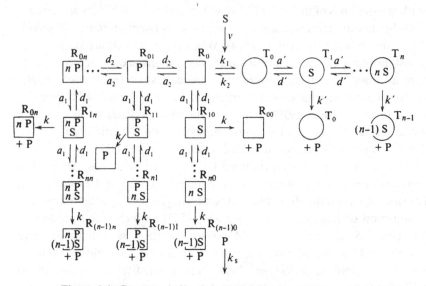

Figure 8.1. Concerted allosteric model for the oscillatory phosphofructo-kinase (PFK) reaction. The enzyme contains n protomers, each of which possesses a catalytic site for the substrate S and a regulatory site for the product P which is a positive effector (see text for details).

could also be obtained with the sequential mechanism proposed by Koshland, Némethy & Filmer (1966). Besides positive feedback, the existence of cooperative interactions between the enzyme subunits is necessary for a nonequilibrium instability in the PFK reaction. This cooperativity may arise from either a concerted or a sequential transition of the enzyme protomers between multiple conformational states upon binding of the substrate or of the positive effector.

The model is represented in Figure 8 1 in the general case of an enzyme containing n protomers. Basic assumptions are as follows.

   (i) The substrate is injected into the system at a constant rate v.

   (ii) Each protomer can exist in two states, R and T, which differ in their affinity for the substrate (K effect) and/or in their catalytic activity (V effect). The protomers undergo a concerted transition between the forms R_0 and T_0 which are free of ligands.

   (iii) The product of the reaction binds exclusively to the R state, which has the largest affinity for the substrate and/or the largest catalytic activity.

   (iv) The reaction product leaves the system at a rate proportional to its concentration (this amounts to the assumption of a nonsaturated Michaelian sink).

In the following, we shall treat the situation of a system that is homogeneous in space; this situation holds in the experiments carried out in continuously stirred extracts of yeast or muscle. A review of the theoretical results on the occurrence of space-dependent structures in the PFK allosteric model has been given by Goldbeter & Nicolis (1976). In homogeneous conditions, the time evolution of the metabolite and enzyme concentrations is governed by the following differential equations (see also Dalziel, 1968):

$$dR_0/dt = -k_1 R_0 + k_2 T_0 - n a_2 P R_0 + d_2 R_{01} - n a_1 S R_0 + (d_1 + k) R_{10},$$

$$\vdots$$

$$dR_{0n}/dt = a_2 P R_{0(n-1)} - n\, d_2 R_{0n} - n a_1 S R_{0n} + (d_1 + k) R_{1n},$$

$$\vdots$$

$$dR_{nn}/dt = a_1 S R_{(n-1)n} - n(d_1 + k) R_{nn},$$

$$dT_0/dt = k_1 R_0 - k_2 T_0 - n a' S T_0 + (d' + k') T_1,$$

$$\vdots$$

$$dT_n/dt = a' S T_{n-1} - n(d' + k') T_n,$$

$$dS/dt = v - n a_1 S \Sigma_0 - (n-1) a_1 S \Sigma_1 - \ldots$$

$$- a_1 S \Sigma_{n-1} + d_1 \Sigma_1 + 2 d_1 \Sigma_2 + \ldots$$

$$+ n d_1 \Sigma_n - n a' S T_0 - (n-1) a' S T_1 - \ldots$$

$$- a' S T_{n-1} + d' T_1 + 2 d' T_2 + \ldots$$

$$+ n d' T_n,$$

$$dP/dt = -n a_2 P R_0 - (n-1) a_2 P R_{01} - \ldots$$

$$- a_2 P R_{0(n-1)} + d_2 R_{01} + 2 d_2 R_{02} + \ldots$$

$$+ n d_2 R_{0n} + k \Sigma_1 + 2 k \Sigma_2 + \ldots$$

$$+ n k \Sigma_n + k' T_1 + 2 k' T_2 + \ldots + n k' T_n$$

$$- k_s P, \tag{1}$$

with the conservation relation

$$R_0 + R_{ij} + T_0 + T_i = D_0 \qquad (i, j = 1, \ldots, n). \tag{2}$$

We have denoted by S and P the concentrations of substrate and product, and by R_{ij} the concentration of the enzymatic form in the R state carrying i molecules of S and j molecules of P; T_i refers to the form in the T state carying i molecules of S. Moreover,

$$\Sigma_i = \sum_{j=0}^{n} R_{ij} \qquad (i = 0, \ldots, n).$$

The various rate constants are defined in Figure 1.

When the concentration of the enzyme is smaller than that of the metabolites, the former varies on a faster time scale. A quasi-steady state hypothesis (Chapter 1) can then be made for the enzyme (Reich & Sel'kov, 1974). Let us define the dimensionless concentrations

$$\alpha = S/K_R, \qquad \gamma = P/K_P, \tag{3}$$

with

$$K_R = d_1/a_1, \qquad K_P = d_2/a_2. \tag{4}$$

The algebraic equations $\dot{R}_0 = 0$, $\dot{R}_{ij} = 0$, $\dot{T}_i = 0$ yield the following expressions relating the enzymatic forms to the dimensionless concentrations of substrate (α) and product (γ):

$$\Sigma_0 = R_0(1 + \gamma)^n, \quad \Sigma_1 = n\alpha e \Sigma_0, \quad \ldots, \quad \Sigma_n = (\alpha e)^n \Sigma_0.$$

$$T_0 = LR_0, \quad \ldots, \quad T_n = (\alpha c e')^n T_0,$$

with

$$R_0 = D_0/[L(1 + \alpha c e')^n + (1 + \alpha e)^n(1 + \gamma)^n]. \tag{5}$$

Insertion of these relations into the kinetic equations for the metabolites yields the final equations:

$$d\alpha/dt = \sigma_1 - \sigma_M \Phi, \tag{6}$$
$$d\gamma/dt = k_s(\lambda \Phi - \gamma),$$

with $\lambda = (q\sigma_M/k_s)$ and

$$\Phi = [\alpha e(1 + \alpha e)^{n-1}(1 + \gamma)^n + L\theta\alpha c e'(1 + \alpha c e')^{n-1}]/[L(1 + \alpha c e')^n$$

$$+ (1 + \alpha e)^n(1 + \gamma)^n]. \tag{7}$$

The parameters σ_1 and σ_M appearing in (6) are the normalized input of substrate and maximum enzyme reaction rate:

$$\sigma_1 = v/K_R, \qquad \sigma_M = nkD_0/K_R = V_M/K_R. \tag{8}$$

Furthermore, $q = K_S/K_P$; $e = (1 + \varepsilon)^{-1}$ and $e' = (1 + \varepsilon')^{-1}$ where $\varepsilon = k/d$ and $\varepsilon' = k'/d'$;

$$\theta = k'/k; \qquad c = K_R/K_T, \text{ with } K_T = d'/a'; \qquad L = k_1/k_2.$$

Parameters c and L are, respectively, the nonexclusive binding coefficient of the substrate and the allosteric constant of the enzyme (Monod *et al.*, 1965). The latter is a *perfect K system* when $\theta = 1$, i.e. when the T and R states have the same catalytic activity and differ only by their affinity for the substrate ($c < 1$). When $c = 1$ and $\theta < 1$, the enzyme represents a *perfect V system*. Both parameters L and c play a prominent role in determining the degree of cooperativity of the enzyme response to changes in metabolite concentrations.

Sustained oscillations and excitability in the PFK reaction

To analyze sustained oscillations and excitability, it is particularly convenient to examine the dynamics of the system governed by (6) in the phase plane (α, γ) determined by the substrate and product concentrations. Similar phase-plane analyses have been conducted in previous studies of excitable systems (Fitzhugh, 1961; Hahn *et al.*, 1974; Tyson, 1977). In the phase plane, two curves are of primary importance: these are the *nullclines* $\sigma_1 = \sigma_M \Phi$ and $\gamma = \lambda \Phi$ corresponding to $(d\alpha/dt) = 0$ and $(d\gamma/dt) = 0$, respectively (see Figure 8.2).

A necessary condition for excitable and oscillatory behavior is that the nullcline $\gamma = \lambda \Phi$ should be an S-shaped sigmoid (see below). This happens in a well-defined parameter range, i.e. for large values of L, and above a critical value of λ (T. Erneux & A. Goldbeter, unpublished). Then the curve $\gamma = \lambda \Phi$ passes successively through a maximum and a minimum in α as γ increases (Figure 8.2*a*). In the absence of regulatory feedback, the nullcline $\gamma = \lambda \Phi$ can never be S-shaped (Figure 8.2*b*).

The intersection of the two nullclines defines the steady state (α_0, γ_0); for the present model, this steady state is always unique. The stability properties of the steady state depend on its location on the sigmoid nullcline. Stability is always ensured when the steady state lies to the left of the maximum or to the right of the minimum on the sigmoid (Goldbeter & Erneux, 1978). This can be easily seen by means of an analysis in which the time evolution of infinitesimal perturbations around the steady state is determined.

Let us denote the perturbations in α and γ by x and y, respectively, so that

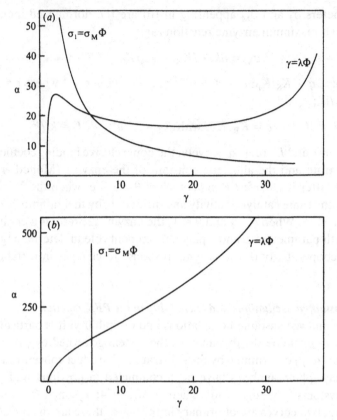

Figure 8.2. The two nullclines of system (6) (*a*) with and (*b*) without positive feedback. Parameter values are $n = 2$, $q = \theta = 1$, $k_s = 0.1\,\text{s}^{-1}$, $\sigma_1 = 0.6\,\text{s}^{-1}$, $\sigma_M = 4\,\text{s}^{-1}$, $L = 10^5$, $c = 10^{-5}$, $\varepsilon = \varepsilon' = 0.1$.

$\alpha = \alpha_0 + x$, $\beta = \beta_0 + y$. Substitution of these equations into (6) and linearization of (6) yields

$$dx/dt = -\sigma_M(\partial\Phi/\partial\alpha)_0 x - \sigma_M(\partial\Phi/\partial\gamma)_0 y,$$

$$dy/dt = k_s\lambda(\partial\Phi/\partial\alpha)_0 x + k_s[\lambda(\partial\Phi/\partial\gamma)_0 - 1]y,$$

(9)

where the subscript zero refers to the steady state. The system (9) being linear admits solutions of the form $x = a\exp(\omega t)$ and $y = b\exp(\omega t)$. Substituting back into (9) yields a homogeneous algebraic system of first degree for a and b. For this system to admit nontrivial solutions requires that its determinant be zero. This condition yields the characteristic equation

$$\omega^2 + \omega[\sigma_M(\partial\Phi/\partial\alpha)_0 + k_s - k_s\lambda(\partial\Phi/\partial\gamma)_0] + \sigma_M(\partial\Phi/\partial\alpha)_0 = 0. \quad (10)$$

Determining the stability properties of the steady state then reduces to analyzing the real part of the solutions of (10).

In the absence of substrate inhibition, i.e. when $c \leqslant 1$, the independent term of the characteristic equation is always positive. Then the condition for ω to have a positive real part is

$$\sigma_M(\partial\Phi/\partial\alpha)_0 + k_s[1 - \lambda(\partial\Phi/\partial\gamma)_0] < 0, \tag{11}$$

or

$$k_s > \sigma_M(\partial\Phi/\partial\alpha)_0[\lambda(\partial\Phi/\partial\gamma)_0 - 1]^{-1}. \tag{12}$$

To relate the condition of instability (12) to the position of the steady state on the sigmoid nullcline, we note that derivation with respect to γ of the relation $\gamma = \lambda\Phi$ yields

$$1 = \lambda[(\partial\Phi/\partial\alpha)(d\alpha/d\gamma) + (\partial\Phi/\partial\gamma)]. \tag{13}$$

Hence,

$$d\alpha/d\gamma = -[\lambda(\partial\Phi/\partial\gamma) - 1]/\lambda(\partial\Phi/\partial\alpha). \tag{14}$$

The relation (14) holds on $\gamma = \lambda\Phi$, and in particular at the intersection point of the two nullclines. Again using the subscript zero to denote this point, we insert (14) into (11) to obtain, as a necessary and sufficient condition for instability,

$$(d\alpha/d\gamma)_0 < -(1/q). \tag{15}$$

It is now clear that the steady state is stable on the ascending branches of the S-shaped curve, where $d\alpha/d\gamma$ is positive, and that unstable steady states, if any, will lie on sufficiently steeply sloping portions of the descending branch. The precise domain of instability can be determined by obtaining the explicit form of the partial derivatives $(\partial\Phi/\partial\alpha)_0$ and $(\partial\Phi/\partial\gamma)_0$ and by the subsequent evaluation of condition (12) on a digital computer (Goldbeter & Lefever, 1972; Goldbeter & Nicolis, 1976). As indicated by (15), the boundaries of the unstable domain tend to coincide with the extrema on the sigmoid as the value of parameter q increases.

Sustained oscillations.   When condition (12) is satisfied, the system governed by (6) undergoes sustained oscillations around the unstable stationary state. In the phase plane, these oscillations correspond to a unique closed trajectory, the limit cycle, which is reached regardless of initial conditions

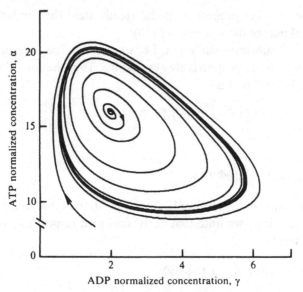

Figure 8.3. Limit cycle in the allosteric model for the PFK reaction. The curve is obtained by numerical integration of (6) for $n = 2$, $\sigma_1 = 0.2 \, \text{s}^{-1}$, $\sigma_M = 10^3 \, \text{s}^{-1}$, $k_s = 0.1 \, \text{s}^{-1}$, $q = 1$, $L = 7.5 \cdot 10^6$, $c = 10^{-2}$, $\varepsilon = 0.1$, $\varepsilon' = \theta = 0$ (the enzyme represents a K–V system); α and γ are the concentrations of ATP and ADP divided by the dissociation constant $K_R = 5 \times 10^{-2} \, \text{mmole l}^{-1}$. The limit cycle can be reached from the outside, or from the unstable steady state ($\alpha = 16$, $\gamma = 2$). The period of sustained oscillations is 145 s (from Goldbeter & Caplan, 1976).

(Figure 8.3). The existence of a limit cycle can be proved mathematically by the construction in the phase plane of a closed domain enclosing the unstable steady state across the boundary of which solutions always head inward (Erle, Mayer & Plesser, 1979).

Experimental data for the parameters of the PFK reaction are available for the enzyme from *Escherichia coli* (Blangy, Buc & Monod, 1968). Inserting the values obtained for the bacterial enzyme in (6), one obtains oscillations whose period is of the order of several minutes. This range agrees with that observed for the oscillations in yeast or muscle. The large values of the allosteric constant L (of the order of 10^5–10^6) used in the simulations are taken from the *E. coli* data; for yeast, such values reflect the existence of regulatory sites for the inhibition of PFK by the substrate ATP (Laurent *et al.*, 1978).

One of the main results obtained in the experiments with yeast extracts has been the demonstration that glycolytic oscillations occur in a finite domain of substrate injection rates comprised between 20 and 160 mmole $\text{l}^{-1} \text{h}^{-1}$ (Hess *et al.*, 1969). This result is easily explained by discussion of the phase portrait in Figure 8.2*a*. As shown above, the domain of sustained oscillations around an unstable steady state lies on the region of negative slope on the sigmoid nullcline $\gamma = \lambda\Phi$. Moreover, the location of the steady state on this curve as a function of the substrate injection rate σ_1 is known, since at the steady state the equations of (6) yield the value $\gamma_0 = q\sigma_1/k_s$ for the product concentration. For low values of σ_1, the steady state lies on the left branch of the sigmoid and is stable. An increase in the substrate injection rate brings the steady state into the region of negative slope; above a critical value of σ_1 the steady state then becomes unstable and limit cycle oscillations develop. Further increase in σ_1 brings the steady state across the instability domain into the right limb of the sigmoid. There thus exists a second, larger critical value of the substrate injection rate above which sustained oscillations disappear and the system evolves toward a stable stationary state. In the model, the oscillatory range of substrate injection rates ranges, typically, from 19 to 246 mmole $l^{-1} h^{-1}$ (Boiteux, Goldbeter & Hess, 1975).

The theoretical characteristics of the oscillations as a function of the substrate input compare with experimental observations both qualitatively and quantitatively for the variation in period and amplitude (Boiteux *et al.*, 1975; Goldbeter & Nicolis, 1976). The model does not account, however, for the double periodicity which is sometimes observed at low substrate injection rates in cellular extracts (Hess *et al.*, 1969). The activity of PFK in the middle of the oscillatory domain determined in the model undergoes a periodic on–off variation between 1% and 75% of the maximum reaction rate, with a mean activity over a period of close to 17% V_M (Figure 8.4); these data match those obtained in yeast extracts (Hess *et al.*, 1969; Boiteux *et al.*, 1975).

Phase shift experiments have shown that the addition of adenine nucleotides at certain times over the period causes a phase advance or a phase delay of glycolytic oscillations (Pye, 1969). Theoretical experiments carried out on a digital computer show, in agreement with experimental observations, that the addition of 0.7 mmole l^{-1} ADP at the minimum of ADP oscillations induces a delay of 1 to 2 min (Goldbeter & Nicolis, 1976). The fact that the addition of ADP has a much smaller influence at other times over the period is easily explicable: the positive effector can appreciably shift the equilibrium between the two enzyme conformations from the T to the R state only when it is added at a minimum of ADP, i.e. when the enzyme is predominantly in the T state and is thus most sensitive to activation by the product.

Entrainment of glycolytic oscillations by a periodic source of substrate has been demonstrated both in experiments with yeast extracts and in the model

(Boiteux *et al.*, 1975; Goldbeter & Nicolis, 1976). Two kinds of entrainment are possible: the enzyme can be locked either to the fundamental frequency of the external input or to a subharmonic frequency. The latter type of entrainment is of special interest as it demonstrates the nonlinear nature of the glycolytic oscillator. It is also reminiscent of the frequency demultiplication that takes place in circadian rhythms which can be altered from a 25 h to a 24 h period by an oscillatory input of 6 h period. Finally, a double periodicity occurs both in the model and the experiments when the period of the external input is much larger than that of the oscillatory enzyme.

Most preceding data have been obtained for a dimeric allosteric enzyme. Phosphofructokinase from many sources is known to be a tetramer (Mansour, 1972). In yeast, experimental studies show that the enzyme contains two types of subunits, probably catalytic and regulatory (Laurent *et al.*, 1978); the number of subunits of each type is either three (Laurent *et al.*, 1978) or four (Tamaki & Hess, 1975; Hofmann, 1978). The yeast enzyme would thus appear as a trimer or a tetramer with dimer subunits. In view of this observation, it is interesting to know the effect of an increasing number of enzyme subunits on metabolic oscillations.

The number of protomers constituting the oscillatory enzyme either has or does not have significant effect on the periodicity, depending on whether or not there exists two distinct time scales in the system of (6). Two time scales will exist for large values of k_s at constant λ, i.e. for large values of the

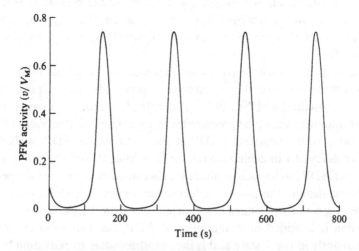

Figure 8.4. Periodic variation on PFK activity. The curve is obtained by integration of (6) on a digital computer for $\sigma_1 = 0.7\,\text{s}^{-1}$, $L = 10^6$ (other parameter values as in Figure 8.2). The ratio v/V_M is equal to function Φ given by (7) (from Goldbeter & Caplan, 1976).

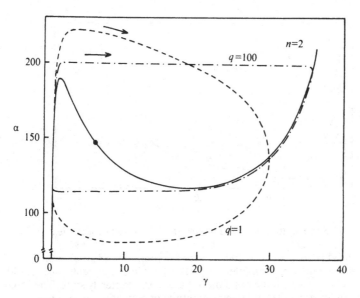

Figure 8.5. Limit cycles for a dimeric enzyme. The closed trajectories
$(---, -\cdot-)$ are obtained at constant λ for $q = 1$, $k_s = 0.1\,s^{-1}$
(period = 366 s) and $q = 100$, $k_s = 10\,s^{-1}$ (period = 186 s); $\sigma_1 = 0.6\,s^{-1}$,
$L = 5 \times 10^6$; other parameters are as in Figure 8.2. The solid line
represents the sigmoid nullcline $\gamma = \lambda\Phi$ on which the dot denotes the
unstable steady state. The value taken for σ_1 yields the largest amplitude
of oscillation for $n = 2$.

parameter q that reflects the differential affinity of the reaction product and
of the substrate for the enzyme. When $q \gg 1$ the amplitude of the oscillations in the product concentration remains practically unchanged as the
number of protomers increases from 2 to 8. This is illustrated in Figures 8.5
and 8.6 where the limit cycles obtained for $n = 2$ and $n = 6$ are shown, with
$q = 100$. In these conditions, the oscillations are of 'relaxation' type,
because the transit times from one branch of the sigmoid to the other (the
horizontal parts on the limit cycle) are short, of the order of several seconds,
whereas the other portions of the limit cycle are covered in minutes. Such
relaxation oscillations give rise to periodic pulses in the product concentration, and have been observed both in a reconstituted glycolytic system
(Hess & Boiteux, 1968) and in intact yeast cells (Von Klitzing & Betz, 1970).

A totally different picture obtains when the two equations of (6) have
similar time scales. Then the amplitude of the oscillations in the product
concentration markedly decreases as the number of protomers passes from 2
to 8. The question arises as to whether this phenomenon is the result of a
reduction in the amplitude of the limit cycle or in the width of the

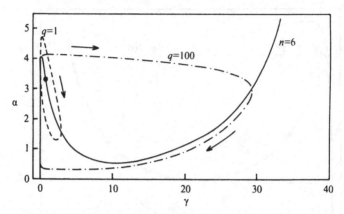

Figure 8.6. Limit cycles for a hexamer. The closed trajectories
($---$, $-\cdot-$) are obtained at constant λ for $q = 1$, $k_s = 0.1\,\text{s}^{-1}$
(period = 78 s) and $q = 100$, $k_s = 10\,\text{s}^{-1}$ (period = 58 s); $\sigma_1 = 0.07\,\text{s}^{-1}$;
other parameters as in Figure 8.5. The solid line represents the sigmoid
nullcline on which the dot indicates the unstable steady state. The value
taken for σ_1 yields the largest amplitude of oscillation for $n = 6$.

sigmoid nullcline as n increases. A phase plane analysis of the effect of n
shows that the former explanation holds, as indicated in Figures 8.5 and 8.6
by the curves drawn for $q = 1$ at constant λ. A further effect of the protomer
number concerns the substrate level and the period, which both diminish as
n increases. The dependence of glycolytic periodicities on the number of
enzyme subunits and on the time scale structure of system (6) has been
analyzed in more detail elsewhere (Venieratos & Goldbeter, 1979; Goldbeter & Venieratos, 1980).

Thinking of the possible physiological significance of metabolic oscillations, one sees that the number of protomers constituting an oscillatory
enzyme can have a decisive influence when metabolic variables in the
oscillatory reaction evolve on similar time scales. Only a low number of
protomers ($n = 2$ to 4) then allows the large amplitude changes in the
product concentration that could initiate a specific metabolic or cellular
response. The above results on the role of n are obtained in systems
regulated by positive feedback, such as the PFK reaction. Different results
are obtained in the analysis of oscillatory enzyme reactions controlled by
end-product inhibition (Walter, 1970).

Excitability. To examine the phenomenon of excitability in the PFK
reaction, it is convenient to resort to a phase plane analysis of the model
governed by (6). As shown in the previous section, the steady state of this

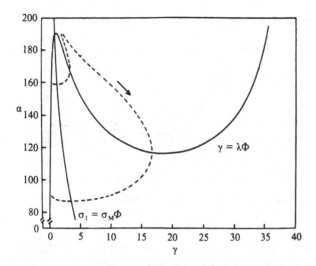

Figure 8.7. Excitability in the phase plane of the PFK model. The stable steady state lies at the intersection of the $\gamma = \lambda\Phi$ and $\sigma_1 = \sigma_M\Phi$ nullclines. Dashed lines represent two trajectories in response to an ADP pulse. The curves are obtained by integration of (6) for $\sigma_1 = 0.1\,s^{-1}$, $L = 5 \times 10^6$ (other parameters are as in Figure 8.2). Initial conditions are $\gamma = 2$ for the small-amplitude trajectory and $\gamma = 2.4$ for the large-amplitude curve; the initial substrate concentration is equal to the steady-state value $\alpha = 189.82$. The excitation threshold is $\gamma = 2$ (see Figure 8.9) (from Goldbeter & Erneux, 1978).

system is stable when located on the left or right branches of the sigmoid nullcline. Any oscillation can only be damped, but the system becomes excitable as it can amplify perturbations whose amplitude exceeds a threshold (Goldbeter & Erneux, 1978). In the situation described in Figure 8.7, the steady state is located just to the left of the maximum on the sigmoid.

Let us consider the response of the system, initially at the stable steady state, to a small instantaneous increase in ADP corresponding to the addition of a pulse of this metabolite. When the amplitude of the pulse is sufficiently low, the system returns immediately to the stable steady state following a small-amplitude trajectory in the (α, γ)-plane (Figure 8.7). When the ADP pulse exceeds a threshold, the system follows a large-amplitude trajectory in the phase plane before returning to the steady state (Figure 8.7). This far-ranging trajectory corresponds to the synthesis of a pulse of ADP followed by a monotone evolution toward the steady state level (Figure 8.8). The above results demonstrate the capability of the enzyme to function as an amplifier of suprathreshold signals, for parameter values close to those for which it generates autonomous oscillations. When

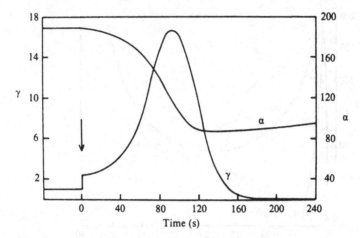

Figure 8.8. Synthesis of an ADP pulse in the PFK reaction. The time evolution of the substrate (α) and product (γ) concentrations is shown in response to a suprathreshold addition of ADP at time zero (arrow). The curves correspond to the large-amplitude trajectory in Figure 8.7 (from Goldbeter & Erneux, 1978).

the parameter values yield a period of several minutes in the oscillatory domain, the time for reaching the maximum response to a signal just above threshold in the near excitable domain is of the order of 100 s (Figure 8.8).

The magnitude of the amplification depends on the parameters of the model and, more specifically, on the value of the allosteric constant. This is well indicated in the dose–response curve in Figure 8.9. Both the amplification and the discontinuous nature of the curve increase with the value of the allosteric constant. Large values of L thus favor excitable behavior just as they favor the onset of sustained oscillations in the PFK reaction (Goldbeter & Lefever, 1972; Goldbeter & Nicolis, 1976). The results of Figure 8.9 demonstrate the possibility of a seven-fold amplification of ADP signals above a threshold. These results are obtained in the absence of any time scale difference between the two equations of (6). As shown in the section devoted to excitability in the adenylate cyclase reaction in *Dictyostelium discoideum*, larger amplification factors and sharper thresholds are obtained when the metabolic variables evolve on different time scales.

The predictions on excitability in the PFK reaction could be checked experimentally in glycolyzing yeast or muscle extracts. As already mentioned, in yeast the technique of a constant rate or substrate injection has shown that sustained glycolytic oscillations occur when the input of fructose or glucose ranges from 20 to 160 mmol l^{-1} h^{-1} (Hess *et al.*, 1969). The model suggests that excitation by ADP pulses should occur for steady states located

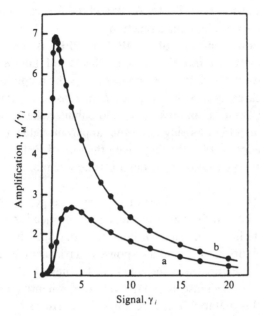

Figure 8.9. Amplification of the ADP signal as a function of stimulation in the PFK reaction. The curves are obtained by successive integrations of (6) for two values of the allosteric constant, $L = 10^6$ (a) and 5×10^6 (b). The amplification is defined as the ratio of the maximum of the synthesized ADP peak, γ_M, divided by the initial ADP concentration, γ_i ($\gamma = 1$ at the steady state). Parameter values are those of Figure 8.7 (from Goldbeter & Erneux, 1978).

to the left of the oscillatory domain in the immediate vicinity of the maximum on the sigmoid nullcline (Figure 8.7). Experimentally, this corresponds to substrate injection rates slightly below $20\,\text{mmole}\,l^{-1}\,h^{-1}$. Using such input rates, the amplification of ADP pulses should be observed above a threshold; the phenomenon should correspond to a burst in NADH fluorescence which is the most suitable marker of oscillating glycolysis. However, the variations in NADH reflect an interplay between the lower and upper parts of the glycolytic pathway (Hess *et al.*, 1969), and it is thus difficult to extract from them quantitative information on a transient activation of PFK. The determination of CO_2 production, which is the end-process of glycolysis, or the assay of metabolites such as fructose diphosphate (FDP) or the adenine nucleotides should bring more clear-cut evidence for excitability in the PFK reaction. It would be useful to interpret variations in the adenine nucleotide pool in the light of a three-enzyme model for glycolytic oscillations proposed by Plesser (1977) who explicitly took into account the activation of PFK by AMP and the formation of the latter metabolite via the adenylate kinase reaction.

When the dissociation constant of ADP for PFK is taken as 5.10^{-5} mole l^{-1} (Blangy *et al.*, 1968), the value predicted in Figure 8.9 for the threshold of stimulation by ADP is 0.1 mmole l^{-1}. As the amplification of the signal becomes larger when the value of L increases, addition of inhibitors of PFK, such as citrate in muscle, should enhance the excitable response, since negative effectors augment the apparent value of the allosteric constant. Besides ADP, other positive effectors of PFK such as FDP in muscle or AMP could be used to demonstrate excitability of the enzyme.

Until now, excitability has been discussed only as a response to the addition of ADP pulses. When the value of the substrate injection rate is such that the steady state lies to the right of the oscillatory domain on the sigmoid nullcline, excitability can occur as a response to a transient decrease in ADP concentration (see Figures 8.10 and 8.11). Such a 'negative' signal is followed first by a further decrease in ADP and, subsequently, by the synthesis of a pulse of this metabolite before the system returns to steady

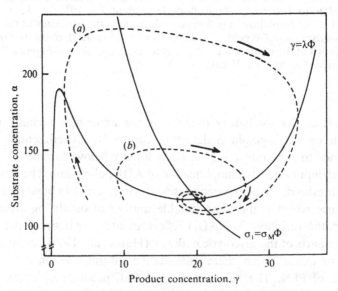

Figure 8.10. Excitation by 'negative' ADP signals in the PFK reaction. The stable steady state ($\alpha = 117.13$, $\gamma = 20$) lies at the intersection of the $\sigma_1 = \sigma_M \Phi$ and $\gamma = \lambda \Phi$ nullclines. The trajectories (dashed lines) follow a 50% (*a*) and a 75% (*b*) reduction in ADP (γ); they are obtained by integration of (6) for $\sigma_1 = 2\,s^{-1}$. Other parameter values are as in Figure 8.7.

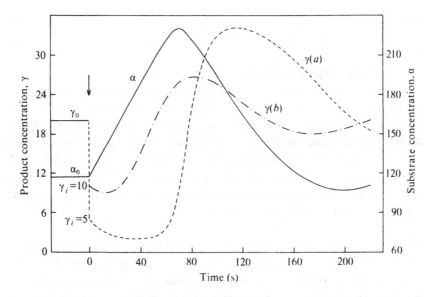

Figure 8.11. Pulsatory response to 'negative' ADP signals in the PFK reaction. The time evolution of the ADP concentration corresponding to trajectories (a) and (b) of Figure 8.10 is shown, with the time evolution of the substrate concentration for (a). α_0 and γ_0 are the steady state levels.

state; a pulse of substrate is also produced in response to the signal. Experimentally, this could be observed in yeast extracts subjected to a substrate injection rate larger than $160 \, \text{mmole} \, l^{-1} \, h^{-1}$, upon a sudden decrease in the activator (ADP or AMP) concentration.

Phosphofructokinase shares the other properties of excitable systems, such as the existence of a refractory period and the increase of the threshold of excitation as the steady state moves further away from the oscillatory domain (Fitzhugh, 1961). These two properties are discussed in greater detail in the following section devoted to the generation of cAMP pulses in *D. discoideum*. Before turning to this system, the question can be asked as to whether all-or-none transitions between multiple steady states are possible in the model for the PFK reaction. Figure 8.2a shows that such a situation can indeed occur in an oscillatory system which loses one degree of freedom: if the substrate level is maintained constant, there exist substrate levels for which the horizontal nullcline $\dot{\alpha} = 0$ will have three intersections with the $\gamma = \lambda \Phi$ sigmoid. Two stable steady states will then be separated by an unstable state. All-or-none transitions will occur from one stable state to the other. These transitions differ from excitable trajectories, since in the latter case the system returns to the original steady state after the pulsatory amplification of the initial perturbation.

More complex modes of oscillatory behavior

Besides excitability and simple periodic oscillations of the limit cycle type examined so far, more complex modes of dynamic behavior can be encountered in regulated enzymatic systems. The origin of more complex dynamic behavior can be investigated by means of enzymatic models based on the product-activated enzyme reaction responsible for glycolytic oscillations. These models depart from experimental observations, but permit one to address in some detail the conditions that favor the transition from simple to complex oscillatory behavior.

Multiple periodic attractors: birhythmicity.   While most oscillatory systems admit a single mode of periodic behavior under given conditions, the question arises as to the possibility of multiple, co-existing oscillatory regimes. The conditions for the occurrence of multiple, stable limit cycles have been investigated in a model based on positive feedback governed by (6), to which a term describing the nonlinear recycling of product into substrate has been added (Moran & Goldbeter, 1984). The introduction of the latter process, which does not modify the number of variables, modifies the shape of the product nullcline in Figures 8.5 and 8.6 in such a manner that a second region of negative slope ($d\alpha/d\gamma$) is obtained. Such a phase portrait allows for the occurrence of two stable limit cycles separated by a third, unstable cycle. This phenomenon, referred to as *birhythmicity* (Decroly & Goldbeter, 1982), permits the system to switch back and forth between two different types of oscillations which co-exist under the same conditions (Moran & Goldbeter, 1984). Another two-variable model in which two allosteric enzymes which share the same substrate and product are coupled in parallel also displays birhythmicity, owing to a similar nullcline structure in the phase plane (Li & Goldbeter, 1989).

Bursting and chaos.   When two allosteric enzymes activated by their respective products are coupled in series, even more complex modes of oscillatory behavior are encountered as a result of the interplay between the two instability-generating mechanisms. In the three-variable model corresponding to such a situation (Decroly & Goldbeter, 1982), in addition to simple periodic oscillations of the limit cycle type as shown in Figure 8.3, birhythmicity is obtained, as well as trirhythmicity, i.e. the co-existence of three simultaneously stable limit cycles. Moreover, the system presents irregular oscillations known as *chaos*: this sort of oscillatory behavior, observed in a number of chemical and biological systems (Olsen & Degn, 1985; Holden, 1986), is characterized by the sensitivity to initial conditions and an apparently random variation in period and amplitude, despite the deterministic nature of the governing equations. In the phase space, a *strange attractor* is reached instead of the limit cycle associated with periodic behavior. Finally, bursting occurs in the multiply regulated biochemical system in the form of high frequency oscillations separated periodically by phases of quiescence. The transition from simple periodic oscillations to patterns of bursting with *n* spikes per period has been investigated in the three-variable enzymatic model (Decroly & Goldbeter, 1987). This detailed analysis reveals the existence of more complex patterns of bursting characterized by different groups of spikes interspersed over a period.

While the above theoretical results on chaos have been obtained for an autonomous system, similar results have been reached for the glycolytic oscillator subjected to an oscillatory input of substrate. Chaotic behavior has been demonstrated theoretically (Markus & Hess, 1984) and experimentally (Markus, Kuschmitz & Hess, 1984, 1985) in these conditions.

The cAMP signaling system in *Dictyostelium discoideum*

In contrast with glycolysis, where the physiological significance of the oscillations remains unclear, the cellular slime mold *D. discoideum* provides an example of an integration of excitable and oscillatory behavior in the life cycle of an organism. *D. discoideum* cells grow as solitary amoebae until they exhaust their food supply. Then, about eight hours after the beginning of starvation, they collect around centers by a chemotactic response to cAMP signals (Konijn, van de Meene, Bonner & Barkley, 1967). As many as 10^5 amoebae may aggregate around a center; the multicellular body thus formed migrates and finally transforms into a fruiting body consisting of only two types of differentiated cells, those that belong to the stalk and those that become spores. This simple pattern of differentiation in an eucaryotic organism, as well as the existence of a mechanism of intercellular communication during the aggregation phase, explains why *D. discoideum* is a major model in developmental biology (Bonner, 1967; Loomis, 1975; Gerisch, 1987).

The process of aggregation in *D. discoideum* has a periodicity of several minutes: waves of inward amoeboid movement appear to propagate outward from the center to the periphery of the aggregation field. To account for this observation, Shaffer (1962) postulated the existence of a dual mechanism for the periodic release of chemotactic attractant by the centers and for the relay of the chemotactic signals by aggregating cells. The chemotactic factor was later identified as cAMP and, subsequently, direct evidence for both the autonomous oscillations of cAMP and the relay of cAMP pulses was obtained in suspensions of *D. discoideum* cells (see below).

The existence of a mechanism for relay and oscillation of the chemotactic signal in *D. discoideum* allows the formation of large aggregation territories (Gerisch, 1968; Cohen & Robertson, 1971; Alcantara & Monk, 1974). Species such as *D. minutum* in which the chemotactic factor is propagated by simple diffusion indeed form much smaller aggregation territories (Gerisch, 1968). In addition to their role as chemotactic signals, periodic cAMP pulses promote and synchronize cell differentiation in *D. discoideum* during the interphase which separates starvation from aggregation (Darmon, Brachet & Pereira da Silva, 1975; Gerisch & Malchow, 1976). This second physiological role consists of inducing the synthesis of specific proteins, most of which belong to the cAMP signal machinery (Klein & Darmon, 1977; Gerisch *et al.*, 1977*a*).

Experimental observations in cell suspensions

The studies of *D. discoideum* suspensions throw light on the cAMP signaling mechanism. Several hours after starvation, the cells exhibit spontaneous oscillations with a period of about seven minutes. These oscillations, revealed by periodic changes in light scattering (Gerisch & Hess, 1974), correspond to sustained oscillations in the concentration of intra- and extracellular cAMP (Gerisch & Wick, 1975). The oscillations can be phase shifted upon addition of cAMP pulses.

The phenomenon of relay is demonstrated in suspensions of cells that have not yet begun – or have ceased – to oscillate. The addition of a pulse of extracellular cAMP (10^{-8} to 10^{-6} mmole l^{-1}) elicits the synthesis of a pulse of intracellular cAMP of much larger magnitude (Roos, Nanjundiah, Malchow & Gerisch, 1975; Shaffer, 1975). The intracellular pulse occurs generally after 1–2 min and is followed by a peak in extracellular cAMP.

Do the phenomena of relay and oscillation originate from a common mechanism or are they caused by two distinct signaling processes? The model analyzed below suggests that the relay of cAMP signals reflects the excitability of adenylate cyclase in *D. discoideum*. It is proposed that the adenylate cyclase reaction in this species of slime mold belongs to the class of excitable and oscillatory chemical systems (Goldbeter, Erneux & Segel, 1978; Martiel & Goldbeter, 1987; Goldbeter, 1990).

Model for the cAMP signaling system

The signaling system consists of a cell surface cAMP receptor and of a functionally coupled adenylate cyclase (Gerisch & Malchow, 1976; Klein, Brachet & Darmon, 1977). The latter enzyme transforms ATP into cAMP. Taking this into account, Goldbeter & Segel (1977) proposed a model based on the observation (Roos & Gerisch, 1976) that binding of extracellular cAMP to the receptor activates adenylate cyclase which faces the inner side of the plasma membrane (Farnham, 1975). This model is a modification and an extension of that proposed by Goldbeter (1975) for cAMP oscillations on the basis of the results of Rossomando & Sussman (1973) for the intracellular regulation of adenylate cyclase. Here, the control of the enzyme by extracellular cAMP is analyzed explicitly.

How the activation of adenylate cyclase proceeds is still unclear. The mechanism might involve allosteric regulation either directly by extracellular cAMP or by an intracellular intermediate whose concentration rapidly changes upon reception of the signal. The enzyme could also be regulated by covalent modification (Greengard, 1978). The observation that a pulse of intracellular cyclic GMP (cGMP) precedes the synthesis of a cAMP pulse (Mato *et al.*, 1977) suggests a possible control of adenylate cyclase by a cAMP-dependent protein kinase (Gerisch *et al.*, 1977*a*). Experimental evidence for such a process is, however, still lacking. Moreover, the cGMP response could be linked to the chemotactic process only, since it is observed in amoeboid species that lack the mechanism for relay and oscillation (Mato & Konijn, 1977). Calcium could be the intracellular effector (Rapp & Berridge, 1977), since a rapid calcium influx follows cAMP binding to the cell surface receptor (Wick, Malchow & Gerisch, 1978). Calcium ions seem, however, to affect the basal activity of adenylate cyclase rather than the oscillations (Roos, Scheidegger & Gerisch, 1977). G-proteins appear to be involved in the step linking binding to the cAMP receptor to activation of adenylate cyclase (Janssens & Haastert, 1987).

To represent the regulation by cAMP in the model (Figure 8.12), the

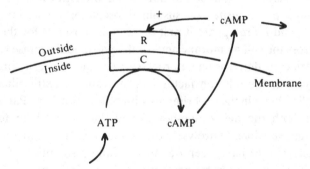

Figure 8.12. Simple model for the cAMP signaling system in *D. discoideum*. R and C denote the cAMP receptor and the functionally coupled adenylate cyclase, respectively. Cyclic AMP (cAMP) is synthesized intracellularly from ATP and transported into the extracelllar medium where it is hydrolysed by phosphodiesterase. The + sign refers to the activation of adenylate cyclase that follows cAMP binding to R (from Goldbeter & Segel, 1977).

simplest assumption is made that the cAMP receptor behaves as a regulatory subunit of adenylate cyclase. Furthermore, the catalytic and regulatory parts of the enzyme are treated as dimers, and the resulting complex is considered as obeying the concerted allosteric model (Monod *et al.*, 1965) with exclusive binding of both substrate and positive effector to the R state. As in the modeling of oscillating glycolysis in yeast extracts, we treat here the case of a spatially homogeneous system. Most experiments on relay and oscillation are indeed performed in continuously stirred suspensions of *D. discoideum* cells. Given the above assumptions, the behavior of the cAMP signaling system is governed by three ordinary differential equations (Goldbeter & Segel, 1977; Goldbeter *et al.*, 1978):

$$d\alpha/dt = v - \sigma\Phi,$$

$$(1/q)\,d\beta/dt = \sigma\Phi - (k_t\beta/q), \tag{16}$$

$$d\gamma/dt = (k_t\beta/h) - k\gamma,$$

where

$$\Phi = \alpha(1 + \alpha)(1 + \gamma)^2/[L + (1 + \alpha)^2(1 + \gamma)^2]. \tag{17}$$

The three metabolic variables α, β and γ denote the concentrations of intracellular ATP, intracellular cAMP and extracellular cAMP divided by K_S, K_P and K_P, respectively, where K_S and K_P are the Michaelis constant of adenylate cyclase for ATP and the dissociation constant of the cAMP receptor. Parameters v and σ relate to the constant ATP input and to the maximum cyclase activity, divided by K_S. (An ATP input is needed to maintain the system far from equilibrium, since the latter condition is the thermodynamic prerequisite for sustained oscillatory behavior.) Also, $q = K_S/K_P$; k_t and k are apparent first order rate constants for the cAMP transport across the cell membrane and for the phosphodiesterase reaction, which are both considered as linear processes; L is the allosteric constant of adenylate cyclase; the dilution factor h is the ratio of extracellular fluid volume to cell volume in the experiments with cell suspensions. Parameter k accounts for both the membrane-bound and the extracellular forms of phosphodiesterase which hydrolyse extracellular cAMP (Farnham, 1975). The equations of (16) have been obtained on the assumption of a quasi-steady state for the enzymatic forms; they are derived in the way outlined for the PFK reaction in the section on glycolytic oscillations.

Sustained oscillations and excitability in the adenylate cyclase reaction

The system (16) admits a single steady state solution whose stability properties can be determined by a linearized stability analysis. To discuss excitable and oscillatory behavior, it is, however, convenient to resort to a phase plane analysis of (16). This is facilitated by the observation that $q \gg 1$. Indeed, $q = K_S/K_P$, where K_S is of the order of $10^{-4} \, \mathrm{mol} \, l^{-1}$ (Klein, 1976; Gerisch & Malchow, 1976) whereas K_P values range from 10^{-9} to $10^{-7} \, \mathrm{mol} \, l^{-1}$ (Henderson, 1975; Gerisch & Malchow, 1976; Mullens & Newell, 1978). The system (16) can thus be approximated by the reduced system (Goldbeter *et al.*, 1978)

$$\beta = (q\sigma/k_t)\Phi, \qquad d\alpha/dt = v - \sigma\Phi, \qquad d\gamma/dt = k(\lambda\Phi - \gamma), \qquad (18)$$

by means of a quasi-steady state assumption for β, in the limit $q \to \infty$ with $\lambda = (q\sigma/hk)$ and (k_t/q) remaining finite. The dynamics of the cAMP signaling system can now be studied in the (α, γ) phase plane.

Sustained oscillations of cAMP. The results obtained in the phase plane analysis of the PFK model hold for the adenylate cyclase system, since the equations of (6) are formally identical to the reduced system (18). As in the case of the PFK reaction, the existence of sustained oscillations is linked to the form of the nullclines $v = \sigma\Phi$ and $\gamma = \lambda\Phi$ and the position of their point of intersection which defines the steady state. A necessary condition for oscillation and excitability is that the nullcline $\gamma = \lambda\Phi$ should be an S-shaped sigmoid.

As in the PFK system, therefore, sustained oscillations around an unstable steady state occur when the intersection of the two nullclines lies in the region of negative slope on the sigmoid $\gamma = \lambda\Phi$ (the instability condition is given by (12) or (15)). Limit cycle oscillations similar to those of Figures 8.3, 8.5 and 8.6 develop, that correspond to the autonomous synthesis of periodic pulses of intra- and extracelllar cAMP (Figure 8.13). These oscillations are accompanied by a periodic variation in ATP, the amplitude of which can be restricted to a 10% variation around the mean ATP level by appropriate parameter choices. Such a reduced amplitude can be related to the observation that ATP remains practically constant in the course of cAMP oscillations (Roos *et al.*, 1977; Geller & Brenner, 1987a). In the model, the oscillations are generally accompanied by variations of larger amplitude in the substrate concentration. A possible compartmentation of the membrane-bound adenylate cyclase reaction *in vivo* could also explain the lack of observable ATP oscillations (see also below, p. 145).

The behavior of the model qualitatively changes when the ATP level is held constant. Oscillations are no longer possible, but the system governed by (18) becomes able to switch between two different stable steady states for a given set of parameter values. Such a situation can be visualized in the (α, γ) phase plane where the nullcline $\dot{\alpha} = 0$ becomes a horizontal line that

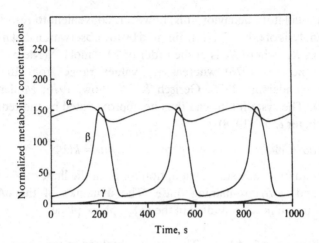

Figure 8.13. Sustained oscillations of intracellular ATP (α), intracellular cAMP (β), and extracellular cAMP (γ). The curves are obtained by integration of the equations of (16) for $v = 0.2\,\text{s}^{-1}$, $k = 1\,\text{s}^{-1}$, $\sigma = 1.2\,\text{s}^{-1}$, $k_t = 0.4\,\text{s}^{-1}$, $q = 100$, $L = 10^6$, $h = 10$.

has one or three intersections with the sigmoid nullcline $\gamma = \lambda\Phi$, depending on the ATP level. The steady state is still unstable when located in the region of negative slope on the sigmoid, but oscillations do not develop as the system evolves toward either one of the two stable steady states. Some variation in ATP is thus needed for oscillatory behavior in the system described by (16) or (18).

Relay of cAMP signals.   The phase plane analysis of the cAMP signaling system is particularly helpful for the comprehension of the relay mechanism, as the response to a pulse of extracellular cAMP can be directly visualized in the (α, γ) plane. When the intersection of the two nullclines is located on the left of maximum A or to the right of minimum D on the sigmoid (see Figure 8.14), the steady state is stable and excitability is mathematically possible. Since relay consists of the amplification of a pulse of extracellular cAMP, the physiological conditions for excitability correspond to a steady state located on the left branch of the sigmoid, near maximum A (Figure 8.14).

In the phase plane, the effect of adding a pulse of extracellular cAMP is simulated by an instantaneous increase in γ, i.e. by a horizontal displacement to the right of the steady state. The terminal point of this displacement defines the initial condition (α_i, γ_i) from which the behavior of the system is determined by integration of (18). The trajectory marked (*a*) in Figure 8.14 shows the response to a suprathreshold stimulation by extracellular cAMP.

The system undergoes a large excursion in the phase plane across the right branch of the sigmoid before returning to the stable steady state. This wide-ranging trajectory in the (α, γ)-plane corresponds to the synthesis of a substantial pulse of intracellular cAMP. As shown by Figure 8.15(a), the relay response obtained by integration of the equations of (16) yields good agreement with the phase plane analysis of the reduced system (18).

The other trajectories in Figure 8.14 indicate that the signaling system can be reexcited on its way back to the steady state. The threshold for these signals decreases and the amplitude of the relay response increases as the system approaches the steady state. The above properties demonstrate the existence of a relative refractory period for relay. During the synthesis of the pulse itself, the refractoriness is absolute and no second signal can be generated on top of the other upon further stimulation. Both types of refractory period have been described by Fitzhugh (1961) in his theoretical study of excitation in nerve membrane.

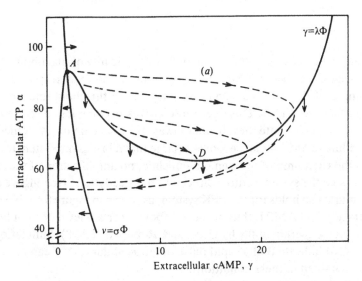

Figure 8.14. Excitability in the adenylate cyclase reaction. In the (α, γ) phase plane of the reduced system (18), the steady state (dot) is located at the intersection of the nullclines $v = \sigma\Phi$ and $\gamma = \lambda\Phi$ on which arrows indicate the local direction of the solution trajectory. Several trajectories (dashed lines) are shown, corresponding to different suprathreshold initial conditions. Trajectory (a) shows the response of the system, initially at steady state, to a pulse of extracellular cAMP. Parameter values are $v = 0.04\,\text{s}^{-1}$, $k = 0.4\,\text{s}^{-1}$; other parameters are as in Figure 8.13. A and D denote a maximum and a minimum, respectively, on the sigmoid (from Goldbeter *et al.*, 1978).

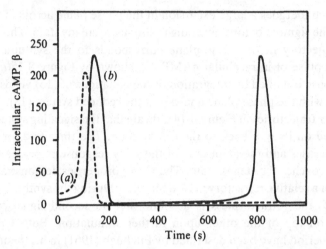

Figure 8.15. Relay of a cAMP signal (*a*) and autonomous oscillations of cAMP (*b*). The curves are obtained by integration of the equations of (16) for $v = 0.04 \text{ s}^{-1}$ (*a*) and $v = 0.1 \text{ s}^{-1}$ (*b*); other parameter values are those of Figure 8.14 (redrawn from Goldbeter & Segel, 1977).

When the steady state lies in the immediate vicinity of maximum *A*, the dose–response curve that links the amplitude of the relay response to the magnitude of the external signal exhibits a sharp threshold followed by a plateau (Figure 8.16*a*). Dose–response curves become less discontinuous when the steady state lies further away from maximum *A*. Once the extracellular cAMP signal becomes so large that it brings the system initially across the right limb of the sigmoid nullcline, no amplification of the signal occurs since the system returns directly to the steady state (see Figure 8.14; this remark also holds for the PFK system, as shown in Figure 8.9). No pulse of extracellular cAMP is thus produced above a certain level of stimulation, as observed experimentally by Grutsch & Robertson (1978). Simulations of system (16) indicate that a rapid pulse of intracellular cAMP can nevertheless be obtained in these conditions.

The theoretical characteristics of relay compare semi-quantitatively with the experimental observations in *D. discoideum* suspensions (Roos *et al.*, 1975; Gerisch *et al.*, 1977*b*), viz. the half-width of the response (of the order of 1 min), amplification factor (of the order of 10–20), and time for maximal response (about 100 s) (Goldbeter & Segel, 1977). The delay of 30–40 s between the peaks in extra- and intracellular cAMP (Roos *et al.*, 1975) can be matched in the model provided phosphodiesterase is treated as a Michaelian enzyme that does not function only in the linear regime (A.

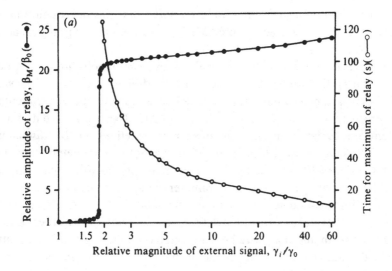

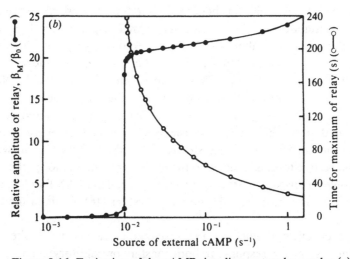

Figure 8.16. Excitation of the cAMP signaling system by a pulse (a) and by a constant source (b) of extracellular cAMP. The amplification factor (●) is given as the maximum of the intracellular cAMP peak, β_M, divided by the steady state level β_0. The second curve in each graph shows the time at which β_M is reached after beginning of stimulation (○). The curves are obtained by integration of the equations of (16) for the parameter values of Figure 8.14. In (b) a constant source term divided by K_P has been inserted in the evolution equation for γ. In (a), the amplitude of the external cAMP signal is given as the initial extracellular cAMP concentration γ_i divided by the steady state level γ_0 which is equal to unity (from Goldbeter & Segel, 1977, and Goldbeter *et al.*, 1978).

Goldbeter, unpublished). It should be noted that this delay could be the result of storage of intracellular cAMP into vesicles prior to transport across the plasma membrane (Maeda & Gerisch, 1977).

The parameter values taken for the above simulations are in part arbitrary. This is the case for L, v and k_t which are difficult to evaluate. The value of these parameters has been adjusted so as to give the experimentally observed steady state levels of ATP and cAMP. Experimental data for the adenylate cyclase and phosphodiesterase reactions in *D. discoideum* (Gerisch & Malchow, 1976; Klein, 1976) suggest that parameters k and σ are of the order of $10^{-2}\,\mathrm{s}^{-1}$, whereas q is in the range 10^3–10^5. Excitability and oscillations still occur for such parameter values, with metabolite levels in the physiological range, but the time scale of both phenomena becomes too slow. Larger values of the adenylate cyclase activity, of the order of those used in our simulations, have been reported (Brenner, 1978). The isolation and subsequent study of the signaling system in *D. discoideum* ghosts or extracts should facilitate its quantitative analysis.

Until now, excitability has been considered in response to cAMP pulses only. A continuous signal of extracellular cAMP can also elicit the synthesis of a pulse of intracellular cAMP. Dose–response curves obtained in the model for such a constant signal resemble those obtained for the response to cAMP pulses (Figure 8.16b). They exhibit both a threshold and a plateau, and the time for maximal relay decreases as the amplitude of the cAMP source increases. What happens if a constant suprathreshold signal is maintained once the relay response is completed? Depending on the stability properties of the new steady state admitted by (16) in which a constant source term is inserted in the evolution equation for γ, the system will evolve to this steady state or further intracellular cAMP pulses will be synthesized, in which case the modified system (16) admits a limit cycle solution.

The theoretical predictions that excitability can occur in the cAMP signaling system as a response to both constant or pulsatory signals above a threshold agree with the observations of Robertson & Drage (1975) on relay in *D. discoideum* fields on agar. The value of the threshold predicted for a pulsatory signal (Figure 16a) is of the order of K_P, i.e. in the range 10^{-9}–10^{-7} mole l^{-1} found experimentally for the dissociation constant of the cAMP receptor (Henderson, 1975; Gerisch & Malchow, 1976; Mullens & Newell, 1978). Such a range for the threshold agrees with that found by Robertson & Drage (1975) on agar (see also Grutsch & Robertson, 1978) as well as with the amplitude of cAMP pulses used in the experiments on relay in cell suspensions (Roos *et al.*, 1975; Shaffer, 1975; Gerisch *et al.*, 1977b). For a constant signal (Figure 16b), the value of the threshold is of the order of $10^{-2} K_P$ s^{-1}, i.e. in the range 10^{-11} to 10^{-9} mole l^{-1} s^{-1} depending on the value taken for K_P. This prediction could be tested experimentally in cell suspension experiments.

Excitability can also occur in the model as a response to a 'negative' cAMP signal when the steady state lies on the right branch of the sigmoid nullcline, near minimum D (see Figure 8.14). Then a transient decrease in extracellular cAMP could result in the synthesis of a peak of intracellular cAMP in the way predicted in Figures 8.10 and 8.11 for the PFK reaction. It is difficult to test this prediction experimentally in the slime mold, since the cAMP signaling system has not yet been isolated in cell ghosts or extracts; until then, it will be difficult to control the location of the steady state on the sigmoid nullcline, in contrast with the situation that prevails for glycolysis in yeast where the oscillatory and excitable domains can be controlled by the technique of constant substrate injection.

The phase plane analysis of the cAMP signaling system yields a qualitative explanation for several observations made on relay in *D. discoideum*, e.g. the existence of a threshold for excitation by extracellular cAMP (Cohen & Robertson, 1971) and the existence of a refractory period (Shaffer, 1962; Gerisch, 1968; Robertson & Drage, 1975). The analysis substantiates the description of the aggregation fields as excitable media (Durston, 1973). According to the model, relay and oscillation of cAMP are two phenomena that are necessarily linked and occur in closely related conditions. The fact that the time for maximal relay in the model varies from 100 to 10 s depending on the magnitude of the constant or pulsatory signal (Figure 8.16) could explain why cells in suspensions relay after 100 s (Roos *et al.*, 1975; Shaffer, 1975) whereas those on agar respond after some 15 s (Cohen & Robertson, 1971; Alcantara & Monk, 1974); the latter might receive a larger signal locally, because of the absence of stirring. The model also explains why the half-width and the waveform of the cAMP peaks during relay and oscillations are similar (see Figure 8.15), as observed experimentally (Gerisch *et al.*, 1977*b*): the phase plane trajectories in both conditions follow proximate paths dictated by the form of the sigmoid nullcline.

In support of a common mechanism is the observation that relay and oscillations in *D. discoideum* have a similar temperature dependence (Gross, Peacey & Trevan, 1976). The observation that 2,4-dinitrophenol can suppress the oscillations by rendering the cells excitable (Geller & Brenner, 1978*b*) does not necessarily imply that the two phenomena are caused by two different mechanisms. 2,4-dinitrophenol affects the ATP level and is likely to modify parameter v in the model. If initially v is such that the nullcline $v = \sigma\Phi$ intersects the sigmoid nullcline in the region of negative slope, i.e. in the oscillatory domain, a decrease in v can shift the steady state to the left of maximum A (see Figure 8.14), thus suppressing the oscillations and bringing the system into the excitability domain.

Developmental control of the signaling system

During the hours that follow starvation, *D. discoideum* amoebae are first able to respond chemotactically to cAMP signals; then they become capable of relaying these signals, before being able to generate autonomously periodic cAMP pulses (Robertson, Drage & Cohen, 1972). Goldbeter & Segel (1977) suggested that such a sequence of developmental transitions can be explained in the model for the signaling system by supposing that some parameters are drifting during interphase. If the system is initially in a state where neither relay nor oscillations occur, such a move in the parameter space can bring the cells first into a region where the system is excitable and later into the domain of autonomous oscillations.

To illustrate in a simple way such a sequence of developmental transitions, let us consider the effect of a variation in the substrate injection rate v on the behavior of the system in the phase plane (α, γ). The steady state value of γ is (qv/hk). Starting from a low value of v that corresponds to a steady state located on the left branch of the sigmoid nullcline (see Figure 8.14), an increase in v brings the steady state near maximum A, i.e. in the domain where excitation occurs in response to a suprathreshold pulse of extracellular cAMP. A further increase in v then brings the system into the region of limit cycle oscillations; if continued, the increase in input rate brings the system into a stable steady state which is no longer excitable.

Experimentally, no evidence exists for a variation in the rate of ATP input to adenylate cyclase during interphase. The transitions discussed as a function of v can also take place in the model as a result of a variation in σ or k, which relate to the activities of adenylate cyclase and phosphodiesterase. Experimental evidence indeed exists for a developmental control of these two enzymes during the hours that follow starvation (Klein, 1976; Gerisch *et al.*, 1977a; Klein & Darmon, 1977). As these changes in enzyme activity take place over a time scale which is much larger than the period of cAMP oscillations, the dependence of the behavior of system (16) on σ and k can be determined assuming that these parameters are constants.

A theoretical path in the adenylate cyclase–phosphodiesterase space can be found that would account for the observed sequence no-relay–relay–oscillations (Goldbeter & Segel, 1980). Qualitatively, the path would traverse four regions in this parameter space, in the sequence *ABCD* (see Figure 8.17). Region *B* is the relay domain which is always located to the right of the oscillatory domain *C*. Cells following this path in the course of development would start, at the onset of starvation, from a state *A* with low

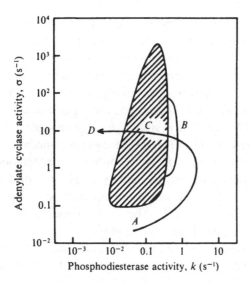

Figure 8.17. Developmental path for the cAMP signaling system in *D. discoideum* as a function of adenylate cyclase and phosphodiesterase activities (see text for details). The oscillatory domain (*C*) and the relay domain (*B*) were determined by linear stability analysis and computer simulation; see text for *A* and *D*. For relay, simulations were performed with the initial condition $\gamma_i = 5$. Parameter values are as in Figures 8.13 and 8.14.

phosphodiesterase and adenylate cyclase activity. This initial state would be devoid of signaling properties and would correspond to a small constant leakage of cAMP into the extracellular medium. Cells entering *B* would become able to relay. Those cells that are the first to enter *C* would become aggregation centers, being able to generate periodic pulses of cAMP. Finally, the cells that leave the oscillatory domain to enter the stable steady state region *D* would secrete cAMP at a high constant rate. This transition, first discussed by Cohen (1977), is thought to occur in the tips of late aggregates. In agreement with experimental observations, the path of Figure 8.17 suggests that cells in such a state *D* are likely to be the cells that previously were aggregation centers in *C*. Other cells could leave the oscillatory domain by returning to *B*, in which case they would again become excitable. Such a process has been observed experimentally in cell suspensions (Gerisch *et al.*, 1977*a*). Let us note here that the behavior of cells in suspensions can differ from the behavior of cells on agar, because the pattern of enzyme synthesis – including that of phosphodiesterase – differs in the two situations (Town & Gross, 1978).

The developmental path suggested in Figure 8.17 fails, however, on two accounts when compared with experimental data. First, the decrease in phosphodiesterase activity required for entering the relay domain prior to the oscillatory regime does not correlate with the time course of this enzyme during aggregation. A decrease in phosphodiesterase activity is indeed observed, but it follows rather than precedes the appearance of autonomous centers. Second, state D in Figure 8.17 does not correspond to a constant level of extracellular cAMP larger than in state A, for a given value of k. This is because, in the model described by (16), the steady state concentration of extracellular cAMP is simply $\gamma_0 = qv/hk$. Thus, contrary to expectations, γ_0 does not depend on the adenylate cyclase activity σ. The reason for this paradoxical result lies in the form of the kinetic equation for α, which at steady state yields the relation $v = \sigma\Phi$. The dependence on σ thereby vanishes at steady state in the equations for β and γ.

In order to remedy this unwanted simplification, one has to modify the evolution equation for ATP. The new system of equations to be considered is

$$d\alpha/dt = v - \sigma\Phi - k'\alpha,$$

$$(1/q)\, d\beta/dt = \sigma\Phi - (k_t\beta/q), \tag{19}$$

$$d\gamma/dt = (k_t\beta/h) - k\gamma,$$

where Φ is given by (17). Comparing with system (16), the only difference is that a term $(-k'\alpha)$ has been added in the kinetic equation for α. As shown below, this single modification overrides the difficulties encountered with the developmental path in Figure 8.17. The motivations for the alteration are twofold. In system (16), the only ATP-consuming process is the reaction catalyzed by adenylate cyclase. This assumption certainly represents an oversimplification since ATP is utilized in other metabolic processes as well. The term $-k'\alpha$ can thus be related to the disappearance of ATP in reactions other than that catalyzed by adenylate cyclase. Alternatively (or simultaneously), this term can be combined with the input rate v and be rewritten as $k'(\alpha_0 - \alpha)$, with $v = k'\alpha_0$. Such a description would correspond to the exchange of ATP between two compartments, one being a constant intracellular pool (α_0), and the other being the layer near the membrane where ATP (α) is immediately available to adenylate cyclase for the synthesis of cAMP. A similar alteration of the kinetic equation for α has been suggested independently by E. L. Coe (personal communication) on the grounds of a comparison between the theoretical and experimentally observed time evolution of ATP.

The dynamic behavior of the model described by (19) in the σ–k plane is shown in Figure 8.18. When k' is less than $10^{-4}\,\mathrm{s}^{-1}$, the effect of the term

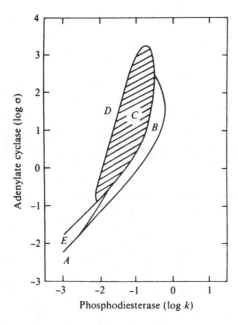

Figure 8.18. Stability diagram in the σ–k plane, for the system governed by eqs (19). Region C denotes the domain in which autonomous oscillations occur around an unstable steady state. In region B, the system is capable of relaying a cAMP signal. This region was determined by computer simulations, taking α and β at steady state and setting the initial concentration of extracellular cAMP as $\gamma = 10$; relay was obtained when the pulse produced in response to the signal exceeded the initial value of γ. Regions A and D refer to stable steady states corresponding, respectively, to low and large levels of extracellular cAMP. In region E, multiple steady states occur (see Figure 8.19). The diagram is established for the parameter values of Figure 8.17, with $k' = 10^{-3}\,\mathrm{s}^{-1}$ (see Goldbeter & Segel, 1980).

$-k'\alpha$ is negligible and the diagram obtained is practically identical to that of Figure 8.17. The diagram of Figure 8.18 has been established for the intermediary value $k' = 10^{-3}\,\mathrm{s}^{-1}$. It shows that the relay domain B has been shifted and lies below the oscillatory domain C. Hence the observed sequence no relay–relay–oscillations can now be obtained in the model without a prior decrease in phosphodiesterase activity, along the path ABC. In addition, the system may leave the oscillatory regime by reentering the relay domain or by going into region D. In the latter case, the steady state level of extracellular cAMP will be larger than at the initial point A since γ_0 now increases with σ as illustrated by Figure 8.19.

A further difference with respect to the graph of Figure 8.17 concerns the

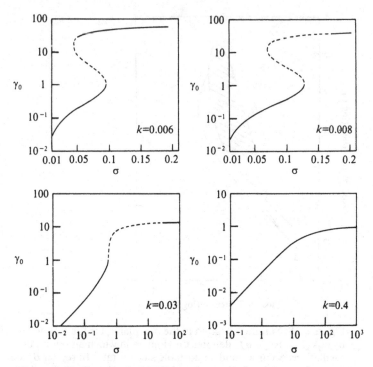

Figure 8.19. Patterns for the dependence of the steady state concentration of extracellular cAMP (γ_0) as a function of the maximum adenylate cyclase activity (σ) obtained for different values of the phosphodiesterase rate constant k in the system governed by equations (19). Multiple steady states obtain for low values of k. A dashed line denotes unstable steady states. The graphs correspond to four different cuts through the diagram of Figure 8.18.

existence of a domain of multiple stationary states (region *E* in Figure 8.18). Two different situations are encountered in this domain, as shown in Figure 8.19. At low values of k, two of the three steady states obtained as a function of σ are stable. For larger values of k, the upper branch of steady states becomes unstable. Let us consider the latter situation, illustrated by the curve for $k = 0.008 \text{ s}^{-1}$ in Figure 8.19. When the value of σ is so large (e.g. $\sigma = 0.15 \text{ s}^{-1}$) that only the upper branch of steady states exists, a limit cycle encloses the steady state when it is unstable. For lower values of σ (e.g. $\sigma = 0.1 \text{ s}^{-1}$) corresponding to three stationary states, the question arises as to whether limit cycle oscillations take place on the upper branch of unstable steady states. Such oscillations have not been found in numerical integrations performed for several initial conditions. If the oscillations exist, starting from another portion of the three-dimensional phase space, a phenomenon of hard excitation could occur on the lower branch of stable steady states. Pulses of extracellular cAMP (γ) could then bring the system from such a steady state to an oscillatory regime around another, higher cAMP level. Such a process could account for the acceleration of development observed upon addition of periodic cAMP pulses (Darmon *et al.*, 1975; Gerisch & Malchow, 1976). Evidence for hard excitation has, however, not been found in the simulations of system (19).

In the situation of one stable and two unstable steady states, numerical integration always showed the system evolving toward the stable steady state. This state is excitable. When perturbed above a threshold value of γ, the system undergoes a large excursion in the phase space, above the upper unstable steady state before returning to the original stable state. Such a behavior gives rise to the synthesis of a cAMP pulse (curve *b* in Figure 8.20).

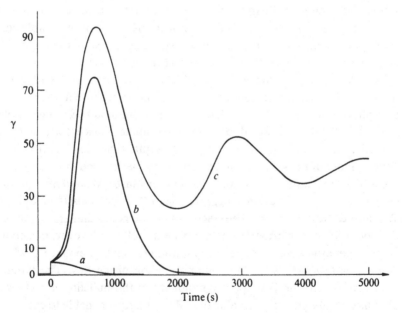

Figure 8.20. Time evolution of the system governed by equations (19) illustrating a transition between two stable steady states (curve *c*, $k = 0.006\,\mathrm{s}^{-1}$, $\sigma = 0.08\,\mathrm{s}^{-1}$), a response to a subthreshold perturbation (curve *a*, $k = 0.006\,\mathrm{s}^{-1}$, $\sigma = 0.05\,\mathrm{s}^{-1}$), and excitation in response to a suprathreshold perturbation (curve *b*, $k = 0.008\,\mathrm{s}^{-1}$, $\sigma = 0.09\,\mathrm{s}^{-1}$). Initial conditions for α and β correspond to steady states located on the lower branch of the curves of multiple steady states in Figure 8.19.
Perturbation of the steady state is realized by increasing at time zero the concentration of extracellular cAMP to the value $\gamma = 5$.

This type of excitable response differs from that encountered in the presence of a single stationary state (Figure 8.14).

When two out of three steady states are stable in region E, all-or-none transitions between them can be observed (curve c in Figure 8.20). For a given value of the maximum adenylate cyclase activity σ, the system can then function with two different effective rates for the enzyme reaction. These two rates correspond to two cAMP steady state levels, one low and the other a hundredfold larger. Upon application of cAMP pulses, the system can leave the lower state and evolve toward the upper branch (Figure 8.20c). Such a behavior would correspond to a transition from an 'inactive' to an 'active' form of adenylate cyclase. Evidence for such a transition in $D.$ *discoideum* amoebae at the beginning of interphase in response to cAMP pulses has been obtained by Juliani & Klein (1978).

The domain of multiple stationary states is small with respect to the oscillatory and relay domains in Figure 8.18. In the course of development, the signaling system may or may not cross the region of multiple steady states, depending on the initial values of parameters k and σ. If it does, the aforementioned phenomena such as transitions between two stable steady states, excitability, or transitions from a stable steady state to an oscillatory regime around a higher mean level of cAMP might occur.

The changes in enzyme activity on the path ABC in Figure 8.18 compare qualitatively with the changes occurring during interphase in the activity of phosphodiesterase and adenylate cyclase (Goldbeter & Segel, 1980). Although the hypothesized path in a parameter space formed by two essential enzyme activities already yields a qualitative explanation for the development of the signaling system, it should be stressed that the real path takes place in a multi-dimensional parameter space. Around fifty genes are essential for the completion of aggregation in $D.$ *discoideum* (Coukell, 1975; Williams & Newell, 1976). The fraction of these genes that are required for the operation of the cAMP signaling system itself is, however, not known.

More generally, any change in any parameter that enters into λ, which is defined as $(q\sigma/hk)$, is likely to alter signaling properties. The latter are indeed linked to the S-shaped sigmoidicity of the nullcline $\gamma = \lambda\Phi$ which obtains only above a critical value of λ. Particularly amenable to experimental manipulation is the dilution parameter h; dilution of the cell suspension can result in the suppression of oscillation and excitability by decreasing λ through an increase in h.

An analogous explanation for the development of the cAMP signaling system has been proposed in a model based on a different regulation of adenylate cyclase (Cohen, 1977). In this, the sequence of developmental changes does not result from a variation in enzyme activities but from a variation, resulting from the decrease of reserve material after starvation, in the level of a catabolite that controls cAMP metabolism. The phase plane dynamics of the cAMP signaling system in that model and in the present one well illustrates how continuous changes in one or two parameters can cause a series of discontinuous developmental transitions.

Model for cAMP oscillations based on receptor desensitization

In the model analyzed above for cAMP oscillations in *Dictyostelium*, substrate consumption plays the role of a counterpoise to self-amplification in cAMP synthesis resulting from the activation of adenylate cyclase that follows cAMP binding to the membrane receptor. However, a number of experimental observations suggest that the primary counterpoise to self-amplification stems from receptor desensitization rather than substrate depletion. Experiments performed with constant cAMP stimuli indicate that cells adapt to such signals (Devreotes & Steck, 1979); moreover, adaptation, i.e. the return to basal cAMP levels despite continuous stimulation by cAMP, can occur in the absence of ATP consumption (Theibert & Devreotes, 1983). Later studies showed that adaptation is brought about by desensitization to cAMP, and that the latter process is associated with phosphorylation of the cAMP receptor (Klein, C., Lubs-Haukeness & Simons, 1985; Devreotes & Sherring, 1985; Vaughan & Devreotes, 1988).

The model for cAMP signalling in *D. discoideum* has been extended to take into account the process of receptor desensitization (Martiel & Goldbeter, 1984, 1987). The number of kinetic equations is then larger, but the system can be reduced to three – and even two – variables, namely, the concentration of extracellular cAMP and the fraction of active (i.e. dephosphorylated) receptor. Then, a periodic variation of the receptor between the active and desensitized forms accompanies cAMP oscillations, as observed in the experiments (Klein, P. *et al.*, 1985). The phase plane analysis of the two-variable version of the model based on receptor desensitization confirms the link between oscillations and excitable behavior, i.e. the relay of suprathreshold pulses of cAMP (Martiel & Goldbeter, 1987). The predictions of the model compare favorably with experimental observations as to the period and amplitude when available parameter values are taken into account.

When taking into account the variation of key parameters such as receptor level, activity of phosphodiesterase and of adenylate cyclase, it is possible to check directly the hypothesis of the developmental path underlying the transitions in cAMP signalling discussed in the previous section. Integration of the kinetic equations of the model based on receptor desensitization then shows that the signalling system switches from no relay to relay, and from relay to oscillations as these parameters increase in a sigmoidal manner during the hours that follow starvation (Goldbeter & Martiel, 1988).

Recently, the two-variable version of this model has been incorporated in a study which successfully accounts for the propagation of cAMP waves in the course of slime mold aggregation on agar (Tyson *et al.*, 1989).

A model similarly based on receptor desensitization, in which the coupling between binding to the receptor and activation of adenylate cyclase is less explicit, has been proposed by Barchilon & Segel (1988). This model, which incorporates the positive feedback exerted by cAMP on its own synthesis via binding to the receptor, also produces relay and oscillations. Another class of model for these phenomena rests on the putative inhibition of adenylate cyclase by calcium ions (Rapp, Monk & Othmer, 1985; Othmer, Monk & Rapp, 1985); this model does not take into account receptor phosphorylation which appears to play a primary role in the mechanism of cAMP oscillations (Gundersen *et al.*, 1989).

Somewhat surprisingly, complex oscillations in the form of bursting, birhythmicity and chaos have been observed in the model for cAMP signaling based on receptor desensitization (Martiel & Goldbeter, 1985; Goldbeter & Martiel, 1985, 1987). The prediction of autonomous chaos could account for the irregular oscillatory behavior observed in the *Fr17* mutant of *Dictyostelium discoideum* which aggregates on agar in an aperiodic manner (Durston, 1974). If this hypothesis proves to be correct, this mutant would provide a first example of autonomous chaos at the cellular level. The analysis of the model indicates that chaos and other complex oscillatory phenomena such as bursting and birhythmicity arise in cAMP synthesis owing to the interplay between two endogenous oscillatory mechanisms coupled in parallel: the two mechanisms share the positive feedback loop exerted by cAMP, and differ by the process limiting self-amplification of cAMP synthesis; in one case, this process is based on substrate consumption, while in the other it relies on receptor desensitization (Martiel & Goldbeter, 1985).

Finally, the role of cAMP signals has been considered in the light of the more general phenomenon of pulsatile intercellular communication. The analysis of models based on receptor desensitization indicates that, as in the case of periodic hormone secretion, there exists an optimal frequency of the pulsatile stimulus that maximizes the responsiveness of target cells (Li & Goldbeter, 1989, 1990). Thus desensitization in target cells provides a mechanism for the efficient encoding of pulsatile signals in terms of their frequency (Goldbeter & Li, 1989; Goldbeter, 1990). Such an encoding may well be one of the most important functions of biochemical rhythms at the cellular level.

Conclusions

Sustained oscillations and excitability are closely associated in chemical systems. The conditions for their occurrence in biochemistry have been studied here in models for two well-known oscillatory systems; glycolysis in yeast or muscle and the cAMP signaling system in *Dictyostelium discoideum*. In both models, the dynamic behavior is governed by a set of nonlinear differential equations which describe the control of an enzyme – phosphofructokinase or adenylate cyclase – by positive feedback. Much information on the dynamics of these reactions can be obtained even when the number of metabolic variables that control the enzymatic system is reduced to only two. The analysis of the models then shows how excitability and sustained oscillations occur in contiguous domains of parameter values. A necessary condition for both phenomena is that one of the two nullclines of the system should be an S-shaped sigmoid in the phase plane determined by the metabolite concentrations.

In the two models, the enzyme is treated as an allosteric protein which obeys a concerted transition mechanism (Monod *et al.*, 1965). Experimental evidence indicates that this assumption holds for phosphofructokinase. In all sources investigated so far – except *D. discoideum* – PFK is an oligomeric protein that exhibits cooperative allosteric interactions (Mansour, 1972). In many instances, the data fit a concerted *K* or *K–V* system, as in *E. coli* (Blangy *et al.*, 1968), muscle (Goldhammer & Hammes, 1978), or yeast (Tamaki & Hess, 1975; Hofmann, 1978; Laurent *et al.*, 1978). The allosteric model for the oscillatory PFK reaction could be extended to take explicitly into account the existence of two substrate–product couples as well as the existence of regulatory sites for the inhibition of the enzyme by ATP. It appears, nevertheless, that the simple monosubstrate allosteric model based on the product activation of PFK already yields agreement with a large number of experiments on glycolytic oscillations in yeast and muscle. In addition, the phase plane analysis of this model shows the possibility of a pulsatory amplification of ADP pulses beyond a threshold. The conditions for observing this phenomenon of excitability in yeast extracts have been determined: the theory predicts that it should occur for substrate injection rates just below those that produce glycolytic oscillations.

In *D. discoideum*, the assumption that adenylate cyclase is an allosteric enzyme still lacks experimental support. The enzyme could in fact be regulated in a different way, e.g. by covalent modification. Thus the question arises as to whether excitability and oscillations are possible in enzyme cascades (Stadtman & Chock, 1978) controlled by positive feedback. Such a situation would apply if adenylate cyclase was regulated by a protein kinase itself under control by cAMP or cGMP (in *D. discoideum* the latter metabolite is synthesized in response to an extracellular cAMP signal prior to adenylate cyclase activation). The analysis suggests that nonequilibrium instabilities could occur in these systems as they occur in reactions controlled by allosteric regulation (Martiel & Goldbeter, 1981). The main reason is the nonlinear nature of the control that cyclic nucleotides exert on protein kinases (Ogez & Segel, 1976).

Another factor that plays a primary role in the mechanism of cAMP oscillations in *Dictyostelium* is the desensitization of the cAMP receptor brought about by covalent modification (Vaughan & Devreotes, 1988; Gundersen *et al.*, 1989). The analysis of a model based on receptor desensitization (Martiel & Goldbeter, 1987) shows that the results obtained by Goldbeter & Segel (1977) hold, provided that the limiting role of substrate consumption is replaced by that of receptor desensitization. In each case, the phase plane analysis of a two-variable version indicates the close link that exists between oscillations and relay of cAMP signals. Different kinetic equations corresponding to different molecular implementations of the autocatalytic synthesis of cAMP thus yield similar results. Therefore, the fact that in *D. discoideum* cAMP itself acts as an extracellular hormone which elicits the synthesis of cAMP suggests that this positive feedback is ultimately responsible, as in many other chemical systems (Nicolis & Prigogine, 1977), for the occurrence of excitable and oscillatory behavior. It should be noted that transient responses to hormonal or chemotactic stimuli, which reflect the existence of an adaptation process, do not generally represent manifestations of excitable behavior. The latter phenomenon is associated with the capability for self-sustained oscillations and implies the existence of a sharp threshold for excitation.

It is the conjunction of cooperativity and feedback regulation that leads to the formation of dissipative structures in biochemical systems (Goldbeter & Nicolis, 1976; Goldbeter & Dupont, 1990). The examples treated here are based on positive feedback. Most models of metabolic oscillations are based on the more frequent control by end-product inhibition (see Walter, 1970; Tyson & Othmer, 1978). Sel'kov (1972) has discussed a variety of regulations that give rise to the phase plane characteristics needed for oscillatory and excitable behavior. Among the oscillatory enzyme reactions presently known, it appears that positive feedback is the mechanism most commonly responsible for the onset of metabolic periodicity (Goldbeter & Caplan, 1976). Besides the regulation by allosteric effectors or by covalent modification, the bell-shaped dependence of enzyme reaction rates upon pH can also lead to instability as a result of autocatalysis in reactions that produce an acid or a base (Hahn *et al.*, 1974). Such is the mechanism of the papain membrane oscillator (Caplan, Naparstek & Zabusky, 1973). The conclusion on the role of positive feedback in the origin of biochemical oscillations is further supported by the analysis of models (Meyer & Stryer, 1988; Dupont & Goldbeter, 1989; Goldbeter, Dupont & Berridge, 1990) that account for the oscillations of cytosolic calcium which have recently been observed in a variety of cells (see Berridge & Galione, 1988, for review).

The sources for nonlinearity in biochemistry are manifold. It is clear, therefore, that regulated biochemical systems frequently possess the capacity for nonequilibrium self-organization of which excitable and oscillatory behavior are two of the most common modes.

References

Alcantara, F. & Monk, M. (1974). Signal propagation during aggregation in the slime mould *Dictyostelium discoideum. J. Gen. Microbiol.* **85**, 321–34.

Barchilon, M. & Segel, L. A. (1988). Adaptation, oscillations and relay in a model for cAMP secretion in cellular slime molds. *J. Theor. Biol.* **133**, 437–46.

Berridge, M. J. & Rapp, P. E. (1979). A comparative survey of the function, mechanism and control of cellular oscillations. *J. Exp. Biol.* **81**, 217–79.

Berridge, M. J. & Galione, A. (1988). Cytosolic calcium oscillators. *FASEB J.* **2**, 3074–82.

Blangy, D., Buc, H. & Monod, J. (1968). Kinetics of the allosteric interactions of phosphofructokinase from *Escherichia coli. J. Mol. Biol.* **31**, 13–35

Boiteux, A., Goldbeter, A. & Hess, B. (1975). Control of oscillating glycolysis of yeast by stochastic, periodic, and steady source of substrate: A model and experimental study. *Proc. Nat. Acad. Sci. USA* **72**, 3829–33.

Bonner, J. T. (1967). *The Cellular Slime Molds.* Princeton, NJ, Princeton University Press.

Brenner, M. (1978). Cyclic AMP levels and turnover during development of the cellular slime mold *Dictyostelium discoideum. Develop. Biol.* **64**, 210–23.

Caplan, S. R., Naparstek, A. & Zabusky, N. J. (1973). Chemical oscillations in a membrane. *Nature, Lond.* **245**, 364–6.

Chance, B., Pye, E. K., Ghosh, A. K. & Hess, B., eds. (1973). *Biological and Biochemical Oscillators.* New York, Academic Press.

Cohen, M. H. & Robertson, A. (1971). Wave propagation in the early stages of aggregation of cellular slime molds. *J. Theoret. Biol.* **31**, 101–18.

Cohen, M. S. (1977). The cyclic AMP control system in the development of *Dictyostelium discoideum. J. Theoret. Biol.* **69**, 57–85.

Coukell, M. B. (1975). Parasexual genetic analysis of aggregation-deficient mutants of *Dictyostelium discoideum. Mol. Gen. Genet.* **142**, 119–35.

Dalziel, K. (1968). A kinetic interpretation of the allosteric model of Monod, Wyman and Changeux. *FEBS Lett.* **1**, 346–8.

Darmon, M., Brachet, P. & Pereira da Silva, L. H. (1975). Chemotactic signals induce cell differentiation in *Dictyostelium discoideum. Proc. Nat. Acad. Sci., USA* **72**, 3163–6.

Decroly, O. & Goldbeter, A. (1982). Birhythmicity, chaos, and other patterns of temporal self-organization in a multiply regulated biochemical system. *Proc. Natl. Acad. Sci., USA* **79**, 6917–21.

Decroly, O. & Goldbeter, A. (1987). From simple to complex oscillatory behaviour: Analysis of bursting in a multiply regulated biochemical system. *J. Theor. Biol.* **124**, 219–50.

De Kepper, P. (1976). Etude d'une réaction chimique périodique. Transitions et excitabilité. *C. R. Hebd. Acad. Sci., Paris, Série C* **283**, 25–8.

Devreotes, P. N. & Sherring, J. A. (1985). Kinetics and concentration dependence of reversible cAMP-induced modification of the surface cAMP receptor in *Dictyostelium. J. Biol. Chem.* **260**, 6378–84.

Devreotes, P. N. & Steck, T. L. (1979). Cyclic 3',5'-AMP relay in *Dictyostelium discoideum.* Requirements for the initiation and termination of the response. *J. Cell Biol.* **80**, 300–9.

Dupont, G. & Goldbeter, A. (1989). Theoretical insights into the origin of signal-induced calcium oscillations. In *Cell to Cell Signalling: From Experiments to Theoretical Models*, ed. A. Goldbeter. London, Academic Press, pp. 461–74.

Durston, A. J. (1973). *Dictyostelium discoideum* aggregation fields as excitable media. *J. Theoret. Biol.* **42**, 483–504.

Durston, A. J. (1974). Pacemaker mutants of *Dictyostelium discoideum. Dev. Biol.* **38**, 308–19.

Erle, D., Mayer, K. H. & Plesser, T. (1979). The existence of stable limit cycles for enzyme catalyzed reactions with positive feedback. *Math. Biosci.* **44**, 191–208.

Farnham, C. J. M. (1975). Cytochemical localization of adenylate cyclase and $3',5'$-nucleotide phosphodiesterase in *Dictyostelium. Exp. Cell Res.* **91**, 36–46.

Fitzhugh, R. (1961). Impulses and physiological states in theoretical models of nerve membrane. *Biophys. J.* **1**, 445–66.

Frenkel, R. (1968). Control of reduced diphosphopyridine nucleotide oscillations in beef heart extracts. 1. Effects of modifiers of phosphofructokinase activity. *Arch Biochem. Biophys.* **125**, 151–6.

Geller, J. & Brenner, M. (1978a). Measurements of metabolites during cAMP oscillations of *Dictyostelium discoideum. J. Cell. Physiol.* **97**, 413–20.

Geller, J. & Brenner, M. (1978b). The effect of 2,4-dinitrophenol on *Dictyostelium discoideum* oscillations. *Biochem. Biophys. Res. Commun.* **81**, 814–21.

Gerisch, G. (1968). Cell aggregation and differentiation in *Dictyostelium*. In *Current Topics in Developmental Biology*, ed. A. A. Moscona and A. Monroy, vol. 3, New York, Academic Press, pp. 157–97.

Gerisch, G. (1987). Cyclic AMP and other signals controlling cell development and differentiation in *Dictyostelium. Annu. Rev. Biochem.* **56**, 853–79.

Gerisch, G. & Hess, B. (1974). Cyclic-AMP controlled oscillations in suspended *Dictyostelium* cells: their relation to morphogenetic cell interactions. *Proc. Nat. Acad. Sci., USA* **71**, 2118–22.

Gerisch, G., Maeda, Y., Malchow, D., Roos, W., Wick, U. & Wurster, B. (1977a). Cyclic AMP signals and the control of cell aggregation in *Dictyostelium discoideum*. In *Developments and Differentiation in the Cellular Slime Moulds*, ed. P. Cappuccinelli andJ. M. Ashworth, Amsterdam, Elsevier/North-Holland Biomedical Press, pp. 105–24.

Gerisch, G., Malchow, D., Roos, W., Wick, U. & Wurster, B. (1977b). Periodic cyclic-AMP signals and membrane differentiation in *Dictyostelium*. In *Cell Interactions in Differentiation*, ed. M. Karkinen-Jääskeläinen, L. Saxén and L. Weiss, New York, Academic Press, pp. 377–88.

Gerisch, G. & Malchow, D. (1976). Cyclic AMP receptors and the control of cell aggregation in *Dictyostelium*. In *Advances in Cyclic Nucleotide Research*, ed. P. Greengard and G. A. Robison, vol. 7, New York, Raven Press, pp. 49–68.

Gerisch, G. & Wick, U. (1975). Intracellular oscillations and release of cyclic AMP from *Dictyostelium* cells. *Biochem. Biophys. Res. Commun.* **65**, 364–70.

Glass, L. & Mackey, M. C. (1988). *From Clocks to Chaos: The Rhythms of Life*. Princeton Univ. Press, Princeton, N.J.

Goldbeter, A. (1975). Mechanism for oscillatory synthesis of cyclic AMP in *Dictyostelium discoideum. Nature, Lond.* **253**, 540–2.

Goldbeter, A. (1990). *Rythmes et Chaos dans les Systèmes Biochimiques et Cellulaires*. Masson, Paris. (English translation to be published by Cambridge University Press, under the title *Rhythms and Chaos in Biochemical and Cellular Systems*.)

Goldbeter, A. & Caplan, S. R. (1976). Oscillatory enzymes. *Ann. Rev. Biophys. Bioengin.* **5**, 449–76.

Goldbeter, A. & Dupont, G. (1990). Allosteric regulation, cooperativity and biochemical oscillations. *Biophys. Chem.* **37**, 341–53.

Goldbeter, A., Dupont, G. & Berridge, M. J. (1990). Minimal model for signal-induced Ca^{2+} oscillations and for their frequency encoding through protein phosphorylation. *Proc. Natl. Acad. Sci. USA* **87**, 1461–5.

Goldbeter, A. & Erneux, T. (1978). Oscillations entretenues et excitabilité dans la réaction de la phosphofructokinase. *C. R. Hebd. Acad. Sci., Paris, Série C* **286**, 63–6.

Goldbeter, A., Erneux, T. & Segel, L. A. (1978). Excitability in the adenylate cyclase reaction in *Dictyostelium discoideum. FEBS Lett.* **89**, 237–41.

Goldbeter, A. & Lefever, R. (1972). Dissipative structures for an allosteric model. Application to glycolytic oscillations. *Biophys. J.* **12**, 1302–15.

Goldbeter, A. & Li, X. Y. (1989). Frequency coding in intercellular communication. In *Cell to Cell Signalling: From Experiments to Theoretical Models*, ed. A. Goldbeter. London, Academic Press, pp. 415–32.

Goldbeter, A. & Martiel, J. L. (1985). Birhythmicity in a model for the cyclic AMP signaling system of the slime mold *Dictyostelium discoideum. FEBS Lett.* **191**, 149–53.

Goldbeter, A. & Martiel, J. L. (1987). Periodic behaviour and chaos in the mechanism of intercellular communication governing aggregation of *Dictyostelium* amoebae. In *Chaos in Biological Systems*, ed. H. Degn, A. V. Holden and L. F. Olsen, New York, Plenum Press, pp. 79–89.

Goldbeter, A. & Martiel, J. L. (1988). Developmental control of a biological rhythm: the onset of cAMP oscillations in *Dictyostelium* cells. In *From Chemical to Biological Organization*, ed. M. Markus, S. Müller and G. Nicolis, Berlin, Springer, pp. 248–54.

Goldbeter, A. & Nicolis, G. (1976). An allosteric enzyme model with positive feedback applied to glycolytic oscillations. In *Progress in Theoretical Biology*, ed. F. Snell and R. Rosen, vol. 4, New York, Academic Press, pp. 65–160.

Goldbeter, A. & Segel, L. A. (1977). Unified mechanism for relay and oscillation of cyclic AMP in *Dictyostelium discoideum. Proc. Nat. Acad. Sci., USA* **74**, 1543–7.

Goldbeter, A. & Segel, L. A. (1980). Control of developmental transitions in the cyclic AMP signalling system of *Dictyostelium discoideum. Differentiation* **17**, 127–35.

Goldbeter, A. & Venieratos, D. (1980). Analysis of the role of enzyme cooperativity in metabolic oscillations. *J. Mol. Biol.* **138**, 137–44.

Goldhammer, A. R. & Hammes, G. G. (1978). Steady-state kinetic study of rabbit muscle phosphofructokinase. *Biochemistry* **17**, 1818–22.

Greengard, P. (1978). Phosphorylated proteins as physiological effectors. *Science* **199**, 146–52.

Gross, J. D., Peacey, M. J. & Trevan, D. J. (1976). Signal emission and signal propagation during early aggregation in *Dictyostelium discoideum. J. Cell Sci.* **22**, 645–56.

Grutsch, J. F. & Robertson, A. (1978). The cAMP signal from *Dictyostelium discoideum* amoebae. *Develop. Biol.* **66**, 285–93.

Gundersen, R. E., Johnson, R., Lilly, P., Pitt, G., Pupillo, M., Sun, T., Vaughan, R. & Devreotes, P. N. (1989). Reversible phosphorylation of G-protein-coupled receptors controls cAMP oscillations in *Dictyostelium*. In *Cell to Cell Signalling: From Experiments to Theoretical Models*, ed. A. Goldbeter, London, Academic Press, pp. 477–88.

Hahn, H. S., Nitzan, A., Ortoleva, P. & Ross, J. (1974). Threshold excitations, relaxation oscillations, and effect of noise in an enzyme reaction. *Proc. Nat. Acad. Sci. USA* **71**, 4067–71.

Henderson, E. J. (1975). The cyclic adenosine 3':5'-monophosphate receptor of *Dictyostelium discoideum. J. Biol. Chem.* **250**, 4730–6.

Hess, B. & Boiteux, A. (1968). Control of glycolysis. In *Regulation of Biological Membranes*, ed. J. Järnefelt, Amsterdam–London–New York, Elsevier Publishing Company, pp. 148–62.

Hess, B. & Boiteux, A. (1971). Oscillatory phenomenal in biochemistry. *A. Rev. Biochem.* **40**, 237–58.

Hess, B., Boiteux, A. & Krüger, J. (1969). Cooperation of glycolytic enzymes. In *Advances in Enzyme Regulation*, vol. 7, Oxford and New York, Pergamon Press, pp. 149–67.

Higgins, J. (1964). A chemical mechanism for oscillation of glycolytic intermediates in yeast cells. *Proc. Nat. Acad. Sci. USA* **51**, 989–94.

Hofmann, E. (1978). Phosphofructokinase – a favourite of enzymologists and of students of metabolic regulation. *Trends in Biochem. Sci.* **3**, 145–7.

Holden, A. (1986). *Chaos*, Manchester Univ. Press.

Janssens, P. M. W. & Van Haastert, P. J. M. (1987). Molecular bases of transmembrane signal transduction in *Dictyostelium discoideum*. *Microbiol. Rev.* **51**, 396–418.

Juliani, M. H. & Klein, C. (1978). A biochemical study of the effect of cAMP pulses on aggregateless mutants of *Dictyostelium discoideum*. *Develop. Biol.* **62**, 162–72.

Klein, C. (1976). Adenylate cyclase activity in *Dictyostelium discoideum* amoebae and its changes during differentiation. *FEBS Lett.* **68**, 125–8.

Klein, C., Brachet, P. & Darmon, M. (1977). Periodic changes in adenylate cyclase and cAMP receptors in *Dictyostelium discoideum*. *FEBS Lett.* **76**, 145–7.

Klein, C. & Darmon, M. (1977). Effects of cyclic AMP pulses on adenylate cyclase and the phosphodiesterase inhibitor of *D. discoideum*. *Nature, Lond.* **268**, 76–8.

Klein, C., Lubs-Haukeness, J. & Simons, S. (1985). cAMP induces a rapid and reversible modification of the chemotactic receptor in *Dictyostelium discoideum*. *J. Cell Biol.* **100**, 715–20.

Klein, P., Theibert, A., Fontana, D. & Devreotes, P. N. (1985). Identification and cyclic-AMP induced modification of the cyclic AMP receptor in *Dictyostelium discoideum*. *J. Biol. Chem.* **260**, 1757–64.

Konijn, T. M., van de Meene, J. G. C., Bonner, J. T. & Barkley, D. S. (1967). The acrasin activity of adenosine-3′,5′-cyclic phosphate. *Proc. Nat. Acad. Sci., USA* **58**, 1152–4.

Koshland, D. E., Némethy, G. & Filmer, D. (1966). Comparison of experimental binding data and theoretical models in proteins containing subunits. *Biochemistry* **5**, 365–85.

Laurent, M., Chaffotte, A. F., Tenu, J. P., Roucous, C. & Seydoux, F. J. (1978). Binding of nucleotides AMP and ATP to yeast phosphofructokinase: Evidence for distinct catalytic and regulatory subunits. *Biochem. Biophys. Res. Commun.* **80**, 646–52.

Li, Y. X. & Goldbeter, A. (1989). Oscillatory isozymes as simplest model for coupled biochemical oscillators. *J. Theor. Biol.* **138**, 149–74.

Li, Y. X. & Goldbeter, A. (1989). Frequency specificity in intercellular communication: The influence of patterns of periodic signaling on target cell responsiveness. *Biophys. J.* **55**, 125–45.

Li, Y. X. & Goldbeter, A. (1990). Frequency encoding of pulsatile signals of cAMP based on receptor desensitization in *Dictyostelium* cells. *J. Theor. Biol.* **146**, 355–67.

Loomis, W. F. (1975). *Dictyostelium discoideum: A Developmental System*, New York, Academic Press.

Maeda, Y. & Gerisch, G. (1977). Vesicle formation in *Dictyostelium discoideum* cells during oscillations of cAMP synthesis and release. *Exp. Cell Res.* **110**, 119–26.

Mansour, T. E. (1972). Phosphofructokinase. In *Current Topics in Cellular Regulation*, ed. B. L. Horecker & E. R. Stadtman, vol. 5, New York and London, Academic Press, pp. 1–46.

Markus, M. & Hess, B. (1984). Transitions between oscillatory modes in a glycolytic model system. *Proc. Natl. Acad. Sci., USA* **81**, 4394–8.

Markus, M., Kuschmitz, D. & Hess, B. (1984). Chaotic dynamics in yeast glycolysis under periodic substrate input flux. *FEBS Lett.* **172**, 235–8.

Markus, M., Kuschmitz, D. & Hess, B. (1985). Properties of strange attractors in yeast glycolysis. *Biophys. Chem.* **22**, 95–105.

Martiel, J. L. & Goldbeter, A. (1981). Metabolic oscillations in biochemical systems controlled by covalent enzyme modification. *Biochimie* **63**, 119–24.

Martiel, J. L. & Goldbeter, A. (1984). Oscillations et relais des signaux d'AMP cyclique chez *Dictyostelium discoideum*: Analyse d'un modèle fondé sur la désensibilisation du récepteur pour l'AMP cyclique. *C.R. Acad. Sci. (Paris) Sér. III* **298**, 549–52.

Martiel, J. L. & Goldbeter, A. (1985). Autonomous chaotic behaviour of the slime mould *Dictyostelium discoideum* predicted by a model for cyclic AMP signalling. *Nature* **313**, 590–2.

Martiel, J. L. & Goldbeter, A. (1987). A model based on receptor desensitization for cyclic AMP signaling in *Dictyostelium* cells. *Biophys. J.* **52**, 807–28.

Mato, J. M. & Konijn, T. M. (1977). Chemotactic signal and cyclic GMP accumulation in *Dictyostelium*. In *Developments and Differentiation in the Cellular Slime Moulds*, ed. P. Cappuccinelli and J. M. Ashworth, Amsterdam, Elsevier/North-Holland Biomedical Press, pp. 93–103.

Mato, J. M., Van Haastert, P. J. M., Krens, F. A., Rhijnsburger, E. H., Dobbe, F. C. P. M. & Konijn, T. M. (1977). Cyclic AMP and folic acid mediated cyclic GMP accumulation in *Dictyostelium discoideum*. *FEBS Lett.* **79**, 331–6.

May, R. M. (1972). Limit cycles in predator–prey communities. *Science* **177**, 900–2.

Meyer, T. & Stryer, L. (1988). Molecular model for receptor-stimulated calcium spiking. *Proc. Natl. Acad. Sci. USA* **85**, 5051–5.

Monod, J., Wyman, J. & Changeux, J. P. (1965). On the nature of allosteric transitions: a plausible model. *J. Mol. Biol.* **12**, 88–118.

Moran, F. & Goldbeter, A. (1984). Onset of birhythmicity in a regulated biochemical system. *Biophys. Chem.* **20**, 149–56.

Mullens, I. A. & Newell, P. C. (1978). cAMP binding to cell surface receptors of *Dictyostelium*. *Differentiation* **10**, 171–6.

Nicolis, G. & Prigogine, I. (1977). *Self-Organization in Nonequilibrium Systems*, New York, Wiley-Interscience.

Ogez, J. R. & Segel, I. H. (1976). Interaction of cyclic adenosine 3′:5′-monophosphate with protein kinase. Equilibrium binding models. *J. Biol. Chem.* **251**, 4551–6.

Olsen, L. F. & Degn, H. (1985). Chaos in biological systems. *Quart. Rev. Biophys.* **18**, 165–225.

Othmer, H. G., Monk, P. B. & Rapp, P. E. (1985). A model for signal-relay adaptation in *Dictyostelium discoideum*. II. Analytical and numerical results. *Math. Biosci.* **77**, 79–139.

Plesser, T. (1977). Dynamic states of allosteric enzymes. In *Proceedings of the VII Internationale Konferenz über Nichtlineare Schwingungen, Abhandlungen der Akademie der Wissenschaften der DDR N6*, vol. 2, Berlin, Akademie-Verlag, pp. 273–80.

Prigogine, I. (1969). Structure, dissipation and life. In *Theoretical Physics and Biology*, ed. M. Marois, Amsterdam, North-Holland Pulishing Company, pp. 23–52.

Pye, E. K. (1969). Biochemical mechanisms underlying the metabolic oscillations in yeast. *Can. J. Bot.* **47**, 271–85.

Rapp, P. E. & Berridge, M. J. (1977). Oscillations in calcium–cyclic AMP control loops form the basis of pacemaker activity and other high frequency biological rhythms. *J. Theoret. Biol.* **66**, 497–525.

Rapp, P. E., Monk, P. B. & Othmer, H. G. (1985). A model for signal-relay adaptation in *Dictyostelium discoideum*. I. Biological processes and the model network. *Math. Biosci.* **77**, 35–78.

Reich, J. G. & Sel'kov, E. E. (1974). Mathematical analysis of metabolic networks. *FEBS Lett.* **40**, S119–27.

Robertson, A. & Drage, D. J. (1975). Stimulation of late interphase *Dictyostelium discoideum* amoebae with an external cyclic AMP signal. *Biophys. J.* **15**, 765–75.

Robertson, A., Drage, D. J. & Cohen, M. H. (1972). Control of aggregation in *Dictyostelium discoideum* by an external periodic pulse of cyclic adenosine monophosphate. *Science* **175**, 333–5.

Roos, W. & Gerisch, G. (1976). Receptor-mediated adenylate cyclase activation in *Dictyostelium discoideum*. *FEBS Lett.* **68**, 170–2.

Roos, W., Nanjundiah, V., Malchow, D. & Gerisch, G. (1975). Amplification of cyclic-AMP signals in aggregating cells of *Dictyostelium discoideum*. *FEBS Lett.* **53**, 139–42.

Roos, W., Scheidegger, C. & Gerisch, G. (1977). Adenylate cyclase activity oscillations as signals for cell aggregation in *Dictyostelium discoideum*. *Nature, Lond.* **266**, 259–61.

Rossomando, E. F. & Sussman, M. (1973). A 5'-adenosine monophosphate-dependent adenylate cyclase and an adenosine 3':5'-cyclic monophosphate-dependent adenosine triphosphate pyrophosphohydrolase in *Dictyostelium discoideum*. *Proc. Nat. Acad. Sci. USA* **70**, 1254–7.

Sel'kov, E. E. (1968). Self-oscillations in glycolysis. 1. A simple kinetic model. *Eur. J. Biochem.* **4**, 79–86.

Sel'kov, E. E. (1972). Nonlinearity of multienzyme systems. In *Analysis and Simulation of biochemical Systems*, ed. H. C. Hemker and B. Hess, Amsterdam, North-Holland Publishing Company, pp. 145–61.

Shaffer, B. (1962). The acrasina. In *Advances in Morphogenesis*, vol. 2, New York, Academic Press, pp. 109–82.

Shaffer, B. M. (1975). Secretion of cyclic AMP induced by cyclic AMP in the cellular slime mould *Dictyostelium discoideum*. *Nature, Lond.* **255**, 549–52.

Stadtman, E. R. & Chock, P. B. (1978). Superiority of interconvertible enzyme cascades in metabolic regulation: analysis of monocyclic systems. *Proc. Nat. Acad. Sci. USA* **74**, 2761–5.

Tamaki, N. & Hess, B. (1975). Purification and properties of phosphofructokinase (EC 2.7.1.11) of *Saccharomyces carlsbergensis*. *Hoppe-Seyler's Zeitschrift für Physiologische Chemie*, **356**, 399–415.

Termonia, Y. & Ross, J. (1981). Oscillations and control features in glycolysis: Numerical analysis of a comprehensive model. *Proc. Natl. Acad. Sci. USA* **78**, 2952–6.

Theibert, A. & Devreotes, P. N. (1983). Cyclic 3',5'-AMP relay in *Dictyostelium discoideum*: adaptation is independent of activation of adenylate cyclase. *J. Cell Biol.* **97**, 173–7.

Tornheim, K. & Lowenstein, J. M. (1974). The purine nucleotide cycle. IV. Interactions with oscillations of the glycolytic pathway in muscle extracts. *J. Biol. Chem.* **249**, 3241–7.

Town, C. & Gross, J. (1978). The role of cyclic nucleotides and cell agglomeration in postaggregative enzyme synthesis in *Dictyostelium discoideum*. *Develop. Biol.* **63**, 412–20.

Tyson, J. J. (1977). Analytic representation of oscillations, excitability, and traveling waves in a realistic model of the Belousov–Zhabotinskii reaction. *J. Chem. Phys.* **66**, 905–15.

Tyson, J. J., Alexander, K. A., Manoranjan, V. S. & Murray, J. D. (1989). Spiral waves of cyclic AMP in a model of slime mold aggregation. *Physica D.* **34**, 193–207.

Tyson, J. J. & Othmer, H. G. (1978). The dynamics of feedback control circuits in biochemical pathways. In *Progress in Theoretical Biology*, ed. F. Snell and R. Rosen, vol. 5, New York, Academic Press, pp. 2–62.

Vaughan A. & Devreotes, P. N. (1988) Ligand-induced phosphorylation of the cAMP receptor from *Dictyostelium*. *J. Biol. Chem.* **263**, 14538–43.

Venieratos, D. & Goldbeter, A. (1979). Allosteric oscillatory enzymes: influence of the number of protomers on metabolic periodicities. *Biochimie* **61**, 1247–56.

Von Klitzing, L. & Betz, A. (1970). Metabolic control in flow systems. 1. Sustained glycolytic oscillations in yeast suspensions under continual substrate infusion. *Archiv. Mikrobiol.* **71**, 220–5.

Walter, C. (1970). The occurrence and the significance of limit cycle behavior in controlled biochemical systems. *J. Theoret. Biol.* **27**, 259–72.

Wick, U., Malchow, D. & Gerisch, G. (1978). Cyclic-AMP stimulated calcium influx into aggregating cells of *Dictyostelium discoideum*. *Cell Biol. Int. Rep.* **2**, 71–9.

Williams, K. L. & Newell, P. C. (1976). A genetic study of aggregation in the cellular slime mould *Dictyostelium discoideum* using complementation analysis. *Genetics* **82**, 287–307.

Winfree, A. T. (1972). Spiral waves of chemical activity. *Science* **175**, 634–6.

Winfree, A. T. (1980). *The Geometry of Biological Time*. New York, Springer.

9

Control of neurotransmitter release: use of facilitation to analyze the regulation of intracellular calcium

Introduction

The subject of this chapter is a quantitative analysis of the synaptic release of transmitter. We shall begin with the essential biological background.

There are basically two ways by which signals are transferred from a nerve to another nerve or to a muscle: one electrical and the other chemical. We will concentrate on the latter.

Chemical signals are transferred through the **synapse**, the region that connects the nerve to the target cell. The organization of a typical synapse is shown in Figure 9.1. The main components of a synapse are (i) the **nerve terminal** whose membrane is the **presynaptic membrane** and in which the **synaptic vesicles** are seen; (ii) the **postsynaptic membrane**, which is the membrane of the target cells; and (iii) a **gap** between the membranes whose thickness is typically somewhat greater than 50 nm. In Figure 9.2, one can see the steps involved in signal transmission as described by Katz (1965). (*a*) The nerve impulse reaches the nerve terminals. (*b*) The impulse causes release of transmitter from the terminals into the gap. (*c*) The transmitter diffuses through the gap to the postsynaptic membrane. (*d*) At the postsynaptic membrane the transmitter binds to receptors. This causes permeability changes in the post-synaptic membrane and thus brings about an **end-plate potential** (e.p.p.).

A typical e.p.p. is seen in Figure 9.3*b*. If this potential reaches a threshold level, it evokes a muscle (nerve) impulse that propagates along the muscle (nerve). The size of the e.p.p., if it is not too high, is the experimental indication of how much transmitter has been released.

Transmitter is released from the nerve terminal very rapidly, once the impulse reaches it. This 'synchronized release' has been shown to *require* external calcium (Del Castillo & Katz, 1954*a*) and to be *antagonized* by

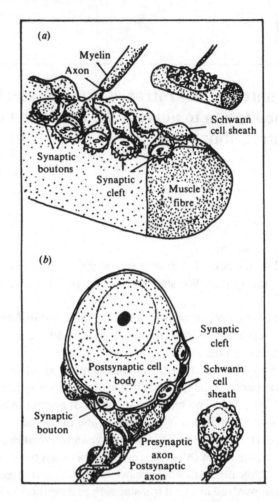

Figure 9.1. Organization of a typical synapse. (*a*) Nerve–muscle synapse; (*b*) nerve–nerve synapse (from Kuffler & Nicholls, 1976).

external magnesium (Katz & Miledi, 1967*a*, *b*). Even if the nerve is not stimulated, end-plate potentials are still seen across the post-synaptic membrane, but they are small and appear spontaneously and randomly. These potentials are called **miniature end-plate potentials** (m.e.p.p.); a typical picture of them is seen in Figure 9.3*a*. The m.e.p.p. do not require external calcium, but their frequency increases greatly after an e.p.p. (Del Castillo & Engbaek, 1954; Miledi & Thies, 1967). This increased frequency is antagonized by external magnesium, in a way similar to the antagonism by Mg of **evoked release** (the release that follows nerve impulse) (Silinsky, Mellow & Phillips, 1977).

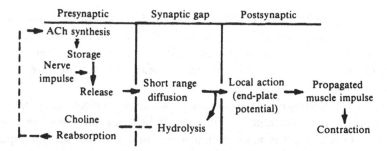

Figure 9.2. Chemical signal transmission (from Katz, 1965).

The m.e.p.p. represents quantal release, probably the content of one vesicle, and the evoked release is built of many such units. The average number of units that is released after an impulse is several hundred, during about 1 ms. Since in frog the mean frequency of m.e.p.p. is $1\,s^{-1}$ (Del Castillo & Katz, 1954b), there is an increase of 10^5 in the frequency of transmitter release after an impulse.

We will concentrate here on evoked release. Two types of measurements are of interest. The first concerns the total (average) number of units that are released after an impulse, the **quantal content**. The other is the measurement of the time course of the released quanta.

Each type of measurement provides a different type of information

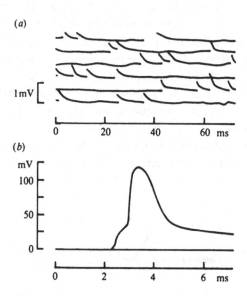

Figure 9.3. (a) A typical miniature end-plate potential (m.e.p.p.); (b) a typical end-plate potential (e.p.p.) (from Kuffler & Nicholls, 1976).

regarding the process of release. A detailed account of the contribution of each of these procedures is given by H. Parnas, I. Parnas & Segel (1990). Here we will concentrate on the role of measuring quantal content in clarifying various fundamental aspects associated with release of neurotransmitter.

It was found many years ago that, if an impulse is given presynaptically, a certain number of quanta are released. If 5, 20 or even 50 ms later a second impulse is given, the amount that is released after the second impulse is normally higher than after the first (Eccles, Katz & Kuffler, 1941). The release continues to increase with successive shocks until it eventually saturates. This increase in the amount of transmitter released with successive impulses is called **facilitation** if one or few impulses are given, and **potentiation** if a train of impulses is given. Typical examples of facilitation are seen in Figure 9.4

The basic theory of facilitation was put forward by Katz & Miledi (1967*b*) who suggested that when an impulse is given, Ca becomes attached to active sites on the membrane and that the number of quanta that are released reflects the number of active sites to which Ca is bound. Ca is continuously removed from those sites. When the second impulse arrives, if there are still some sites occupied by Ca then the new number of occupied sites is higher than after the first impulse and more transmitter will be released. This is the 'residual theory'.

The residual theory was further developed by Dodge & Rahamimoff (1967). They showed that the amount of transmitter being released follows a sigmoid curve as a function of external Ca concentration, the slope of which is approximately 4 when Ca is low. They suggested therefore that only sites which are occupied by four molecules of Ca are effective for release. When the second impulse arrives, if there are sites which have two or three Ca molecules, it will be easier to reach the required 'four-state' and thus more transmitter will be released.

The 'residual theory' that describes accumulation of a Ca complex can easily be extended to describe accumulation of Ca in the nerve terminal if a

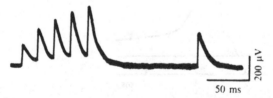

Figure 9.4. Facilitation during and following 5 impulses (from Mallart & Martin, 1967).

connection between release and internal Ca can be shown. Relevant evidence was obtained when Miledi (1973) injected Ca into nerve terminals of a squid and found elevated transmitter release without nerve stimulations and with no external Ca. There are important differences between normal evoked release and the elevated release that followed injection. If one consults Figures 9.5 and 9.3, one can see that normal e.p.p. is both higher and much shorter than the release as a result of injection of Ca. Still Miledi's experiments support the notion that release reflects the level of free Ca in the nerve terminals and that facilitation is the result of accumulation of Ca in the terminals. Further support comes from the work of Alnaes & Rahamimoff (1975) which shows that inhibition of the uptake of Ca into mitochondria increases release. Moreover, Connor *et al.* (1986) showed that potentiation has a similar time course to that of the calcium indicator arsenazo III.

From the above discussion it follows that in order to understand the dependence of release on the internal Ca concentration, one must understand the entry and removal processes. Unfortunately, even with very sophisticated modern techniques to monitor the levels of internal Ca, this goal is difficult to achieve. A major reason is the small size of nerve terminals. Extrapolation from other synapses might distort the real picture. Indirect methods must therefore be employed to study entry, removal and even release itself. A very powerful indirect tool is facilitation (defined below). Since facilitation depends on intracellular Ca, experiments can be designed to unravel features associated with entry, removal or release. To this end, understanding and formulating the dependence of facilitation on Ca is required. In the following we discuss this subject.

One can visualize facilitation and also evoked release as arising from three processes: (*a*) **entry**, (*b*) **removal**, and (*c*) **release**.

(*a*) During nerve stimulation, conductance changes occur in the presynaptic membrane and Ca enters the nerve terminal. Thus the

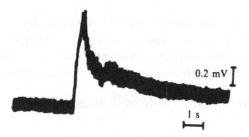

0.2 mV

1 s

Figure 9.5. Transmitter release following injection of Ca into squid nerve terminals (from Miledi, 1973).

level of free Ca is temporarily raised from a resting level to a new higher level.

(b) Internal Ca is continuously removed and the level of free Ca decreases back toward the resting level.

(c) Transmitter is rapidly released following the nerve impulse in an amount that reflects the concentration of internal free Ca.

When two pulses are given, let L_1 and L_2 denote the amounts of transmitter released after the first and second pulses. The facilitation (F) is quantitatively defined by

$$F = L_2/L_1. \tag{1}$$

Experimentally, the decay of facilitation is measured by giving the second (test) pulse after increased durations of time. In this context F is regarded as a function of the time t between the two pulses.

Previous models to describe F

Facilitation has been treated quantitatively by several authors. Some of the equations as well as the assumptions that led to them will now be briefly described.

(i) $F(t) = F_0 \exp(-\alpha t)$,   Mallart & Martin (1967).

Mallart and Martin characterized facilitation as being composed of two components, early and late, according to the time course of their decay. These authors proposed that the magnitude and time course of both components of facilitation are the same for every shock in a short train of repetitive stimulations, and that the individual facilitatory effects sum linearly. They also assumed that both components of facilitation decay exponentially. Equation (i) describes the decay of the early component.

(ii) $F = [1 + \exp(-\alpha t)(1 - A)]^4$,   Rahamimoff (1968).

Rahamimoff defined A as the fraction of sites occupied by Ca and thus obtained the results $L_1 = KA^4$ and $L_2 = KA^4(1 - A\exp(-\alpha t) + \exp(-\alpha t))^4$. He assumed cooperativity in release and also assumed that the removal of Ca from the active sites follows an exponential decay curve.

(iii) $F(t) = F_0 \exp(-t/\tau)$,   Balnave & Gage (1974).

Similarly to (i).

(iv) $f(t) = 0.8 \exp(-t/50) + 0.12 \exp(t/300)$

+ 0.025 exp $(-t/3000)$,   Magleby (1973).

Here $f(t)$ is the facilitation contributed by each impulse. According to Magleby, this facilitation has three components, each of which decays exponentially with a different time course.

(v) $A \xleftrightarrow[k_{-1}]{k_1} B \xleftrightarrow[k_{-2}]{k_2} C$,   Balnave & Gage (1977).

Here the first reaction represents influx of Ca and the second conversion of Ca from B to an activated form C. The decay of C is responsible for the decay of facilitation and this is given by $C = C_0 \exp(-k_2 t)$, again an exponential decay.

Without separately discussing each of the above models, and others like them that have not been mentioned, one can see that some of the models in the literature do not suggest any physical mechanism for the release and facilitation process, others do suggest a mechanism, but all of them postulate that facilitation or its cause *decays exponentially*. Recall now the picture that was described earlier for release, which assumed that release reflects the internal free Ca concentrations. According to this picture an exponential decay of facilitation or its cause means that the internal Ca concentration declines exponentially after being raised temporarily after the pulse.

Exponential decay of the intracellular Ca concentration would imply that the overall rate of removal of the entered Ca is linearly dependent on its concentration. Such a conclusion would suggest that the main process governing the removal of Ca is diffusion, since the rate of other processes (such as extrusion by means of the Na $\leftrightarrow$ Ca exchanger) saturate at higher Ca concentrations. It is therefore of utmost importance to establish the nature of the lumped removal processes that together govern the time course of intracellular Ca concentration. As mentioned earlier, a combined theoretical and experimental analysis of facilitation will serve as a tool for achieving this goal.

Release and facilitation are the outcome of three tightly connected processes, entry of Ca, release of neurotransmitter, and removal of Ca. Hence, understanding one of these processes heavily depends on understanding all three. We will start with the release equation, which is only slightly dependent on the nature of the entry equation and not at all on the removal equation.

Before turning to the details of our theoretical analysis, we shall make some general remarks about notation. As has already been stated, our models will be built of three elements – entry, removal, and release. The italicized letters provide mnemonics for the notation that we shall employ. The letters 'E' and 'ε' will be associated with entry, 'μ' with removal, and 'L'

and 'λ' with release. In particular, when functions of saturation type are assumed, the 'Michaelis constants' that give the concentrations for half-saturation will be denoted by K_ε, K_μ, and K_λ respectively. In connection with these saturation functions, we point out that it is the *general* saturation phenomenon that will prove of central importance; the true saturation functions could well be different from the Michaelis–Menten functions used in our theory, but this would not change the main lines of our argument.

We shall employ the notation $C = C(t)$ to denote the internal Ca concentration at time t, and C_e to denote the external Ca concentration (assumed constant).

The release equation

Dodge & Rahamimoff (1967) found, as was mentioned earlier, that a sigmoid curve relates the amount of transmitter released after an impulse to the external Ca concentration. In particular, this curve is observed to saturate as C_e becomes large. Moreover, Rahamimoff (1968) in the frog and H. Parnas, Dudel & I. Parnas (1982) in the crayfish showed that facilitation was a decreasing function of C_e in experiments where the time interval between the two pulses was very short. We shall now demonstrate that these experimental facts mandate the assumption that release saturates at high internal Ca concentrations.

The observed saturation in release could occur if entry were a saturable function but release itself was linear. Let us work out what facilitation would be in this situation. To do this we first note that entry is expected to be a function of the difference between the external Ca concentration C_e and the internal Ca concentration C (at the time of the impulse). Initially C has a resting value $C_r = 1 \,\mu\text{mole}\,l^{-1}$, a value that is small compared even to the lowest values of C_e normally used in the frog (0.2 mmole l^{-1}). After the first entry, C is still small compared to C_e, so that the second entry will be virtually the same as the first. If entry causes internal Ca concentration to increase by an amount E, then after the first entry $C = C_r + E$. The time between pulses is so short that removal is negligible. After the second entry, then $C = C_r + 2E$. If release is proportional to C, then

$$F = (C_r + 2E)/(C_r + E) \approx 2. \qquad (2)$$

There is no dependence on C_e. Thus the experimental results cannot be explained with the assumption that only entry is a saturable function.

Given the above-mentioned results, it has been shown that we have no alternative to the assumption that release is a saturable function of internal Ca. We will now work out what facilitation would be in this situation.

We first note that the conclusion that release L saturates at high C is best formulated by

$$L(C) = \lambda C^{n_\lambda} / [(K_\lambda + C)^{n_\lambda}]. \tag{3}$$

The power law in (3) is introduced to ensure generality: n_λ could be 1. The point of interest for the present discussion, however, is the effect on experiments in which facilitation is measured of the saturable dependence of L on C.

Let us assume for the present that entry has a linear functional dependence on C_e only,

$$E = \varepsilon(C_e - C) \approx \varepsilon C_e. \tag{4}$$

Remembering that removal is negligible in the experiments under consideration at the moment, we see that the release in the first two pulses is given by

$$L_1 = \lambda \left(\frac{\varepsilon C_e}{K_\lambda + \varepsilon C_e} \right)^{n_\lambda} \quad \text{and} \quad L_2 = \lambda \left(\frac{2\varepsilon C_e}{K_\lambda + 2\varepsilon C_e} \right)^{n_\lambda} \tag{5a, b}$$

so that

$$F = \frac{L_2}{L_1} = \left[\frac{2(K_\lambda + \varepsilon C_e)}{K_\lambda + 2\varepsilon C_e} \right]^{n_\lambda}. \tag{6}$$

Note that $F \to 1$ as $C_e \to \infty$. Moreover it is easy to show that F is a decreasing function of C_e, for

$$\frac{\partial F}{\partial C_e} = n_\lambda \left[\frac{2(K_\lambda + \varepsilon C_e)}{k_\lambda + 2\varepsilon C_e} \right]^{n_\lambda - 1} \frac{\partial}{\partial C_e} \left(\frac{2(K_\lambda + \varepsilon C_e)}{K_\lambda + 2\varepsilon C_e} \right)$$

$$= \frac{-2\varepsilon K_\lambda n_\lambda [2(K_\lambda + \varepsilon C_e)]^{n_\lambda - 1}}{[K_\lambda + 2\varepsilon C_e]^{n_\lambda - 1}} < 0. \tag{7}$$

A summary of the foregoing considerations and the corresponding experimental results are shown in Figure 9.6. This figure demonstrates that the assumption of saturation in release at higher internal Ca concentrations fits the experimental finding of decline in F as C_e increases, when the time interval between pulses is short.

Moreover, (2) and (7) together demonstrate that the reduction in F as C_e

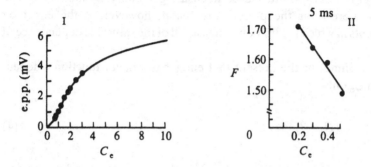

Theoretical analysis

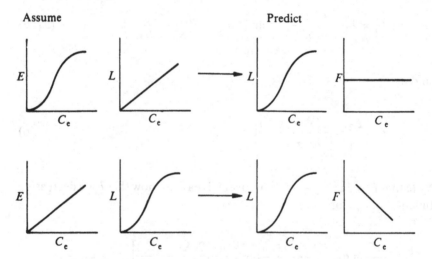

Figure 9.6. Experimental and theoretical considerations in choosing the release equation. *Experimental results*: I. The amount of transmitter released following an impulse relates by a sigmoid curve to the external concentration (from Dodge & Rahamimoff, 1967). II. Facilitation declines as external Ca concentration (C_e) is raised, when the time interval between impulses is 5 ms (from Rahamimoff, 1968). For symbols see text.

Theoretical considerations Saturable entry but linear release predicts only I, while linear entry, but saturable release predicts both I and II.

increases is independent of the nature of the entry process. It thus is possible to extrapolate the saturative dependence of release on C (which cannot be measured) from experiments in which C_e was modified.

The removal equation

F *as a function of* C_e *at short and long intervals between pulses*

The main subject of this portion of the investigation is whether an exponential removal function can account for certain experimental results that concern the relation between facilitation and the external Ca concentration, C_e, and, if not, what the removal function should be. The relevant experimental results are summarized in Figure 9.7a. One can see that when there are short time intervals between pulses then F declines if C_e is raised, as expected from the saturation of the release function L at higher C. On the other hand, when the time interval between pulses is increased to 50 ms, then F *increases* as C_e is raised. This result cannot be predicted on the basis of saturation in release. One can also see in Figure 9.7a_2 that the time interval over which the test pulses still show facilitation (the duration of facilitation) increases as C_e is raised, both in two-pulse experiments (Rahamimoff, 1968; I. Parnas, H. Parnas & Dudel, 1982) and after a train of repetitive stimulation (Rosenthal, 1969). Finally, if the decay of facilitation with time is investigated at two different values of C_e, the curves have an intersection point (Figure 9.7a_3). This result follows from previous observations – at larger values of C_e, facilitation is lower when the test pulse is given shortly after the control pulse, but duration is longer.

Let us consider those experiments for which the interval between pulses is sufficiently long to make it necessary to take removal into account. Using an exponential removal law (8b), we see that (5b) and (6) must be modified to

$$dC/dt = -\mu C, \quad \text{i.e.} \quad C = C_0 \exp(-\mu t). \tag{8a, b}$$

$$L_2 = \lambda \left[\frac{\varepsilon C_e(1 + \exp(-\mu t))}{K_\lambda + \varepsilon C_e(1 + \exp(-\mu t))} \right]^{n_\lambda},$$

$$F = \left[\frac{(1 + \exp(-\mu t))(K_\lambda + \varepsilon C_e)}{K_\lambda + \varepsilon C_e(1 + \exp(-\mu t))} \right]^{n_\lambda}. \tag{9a, b}$$

The quantitative conclusion of (7) remains true, however, for once again $\partial F/\partial C_e < 0$ for all t. That is, in contradiction to experiment, the present theory predicts that F will *always* decrease as C_e is raised, with any time interval between pulses. For time intervals long compared to μ^{-1}, this

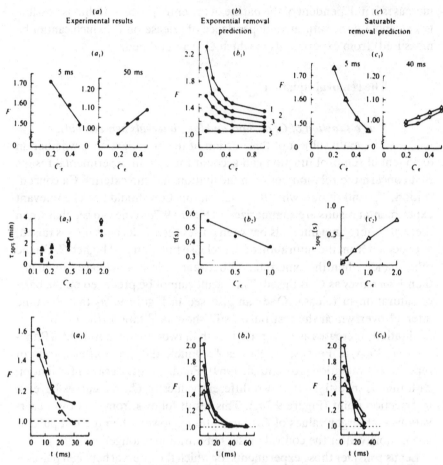

Figure 9.7. Experimental and theoretical considerations in choosing the removal equation. Experimental results (*a*): (*a₁*), facilitation declines as C_e is raised when the time interval between impulses is 5 ms, but increases, as C_e is raised, when the time interval is 50 ms (from Rahamimoff, 1968); (*a₂*), the duration of facilitation increases as C_e is raised following a train of repetitive stimulations (from Rosenthal, 1969); (*a₃*), decay curves of facilitation at ○, $C_e = 0.2$ mmole l^{-1} and ●, $C_e = 0.4$ mmole l^{-1} (Rahamimoff, 1968).

Theoretical considerations. Given saturable functions for entry (equation (13*a*)) and release (equation (3)), and exponential removal (*b*) fails to predict either (*a₁*), (*a₂*), or (*a₃*). See (*b₁*), (*b₂*) and (*b₃*). In (*b₁*) 1–5 are 5, 20, 40, 60 and 80 ms, respectively. In (*b₃*) $C_e = $ ○, 0.1 mmole l^{-1}; ●, 0.2 mmole l^{-1}; △, 0.4 mmole $^{-1}$. Saturable removal (*c*), however, predicts correctly (*a₁*), (*a₂*) and (*a₃*). See (*c₁*), (*c₂*) and (*c₃*). In (*c₁*), △ and ○ are $n_\mu = 1$ and 2. In (*c₃*), ○, ● and △ are $C_e = 0.1$, 0.2 and 0.4 mmole l^{-1}, respectively.

decrease will not be detectable. Graphs of F as a function of C_e at various time intervals between pulses are presented in Figure 9.7.

Duration of facilitation

We provide a quantitative definition for the **duration of facilitation** τ_q as the time that it takes for F to revert to the value q, where q will normally be a few percent above unity. (The letter τ itself will be used in discussions where the precise value of q is immaterial.) From (9b), τ_q can be found as the solution of the equation

$$q^{1/n_\lambda} = \frac{[1 + \exp(-\mu\tau_q)][K_\lambda + \varepsilon C_e]}{K_\lambda + \varepsilon C_e[1 + \exp(-\mu\tau_q)]}. \tag{10}$$

A little manipulation shows that one can write

$$1 + \exp(-\mu\tau_q) = q^{1/n_\lambda}/[1 - \varepsilon C_e K_\lambda^{-1}(q^{1/n_\lambda} - 1)]. \tag{11}$$

It is already clear that an increase in C_e will decrease the denominator on the right side, and therefore will decrease τ_q. Let us exploit the fact that $q \approx 1$, by employing the approximations $(1 - x)^{-1} \approx 1 + x$ and $\ln(1 - x) \approx x$, where x is any quantity satisfying $|x| \ll 1$. (These approximations are just the first terms of the appropriate Taylor series.) By this means, if $q \approx 1$ we can write

$$\tau \approx \mu^{-1} \ln[1/(q^{1/n_\lambda} - 1)] - (\varepsilon C_e/\mu K_\lambda)q^{1/n_\lambda}. \tag{12}$$

This formula explicitly shows that duration decreases as C_e increases, in contrast to the experimental findings depicted in Figure 9.7a_2.

We have demonstrated that the exponential removal function fails to predict correctly *any* of the experimental results which deal with the effect of C_e on facilitation. If one examines the discrepancies between the exponential removal predictions and the experimental results, one can see that the removal process should 'compensate' for the decrease in the derivative of the release function as a function of C_e. Let us demonstrate this assertion in connection with the observation that F is an increasing function of C_e when the time interval between pulses is approximately 50 ms, but is a decreasing function when this interval is considerably less than 50 ms. If 50 ms are enough to remove *all* the Ca that entered at low C_e, but are not enough to remove all the Ca that entered at higher C_e, then $F \approx 1$ at lower C_e and $F > 1$ at higher C_e. Consequently the rate of removal must saturate at higher Ca concentrations, instead of increasing linearly as assumed by the hypothesis of exponential removal. Indeed we will see that this simple assumption of saturation in the rate of removal leads to correct predictions for all the above-mentioned experiments.

It is important to mention that saturation in removal is not only a mathematical requirement which emerged from our theoretical study, but it is also a logical assumption considering the biological facts known about removal. Since it has been pointed out by Alnaes & Rahamimoff (1975) that mitochondria are involved in the removal of internal free Ca, that Ca is extruded through the presynaptic membrane by an active process (Blaustein & Hodgkin, 1969), and that the kinetics follow a Michaelis–Menten curve (Dipolo, 1973), the assumption of saturation seems eminently reasonable. We thus conclude that the rate of removal, dc/dt, is given by (13c) below.

The entry equation

Given that the removal processes show a saturative dependence on C, we can use this characteristic and turn to developing the relation between C_e and the amount of Ca that enters per pulse. Examination of (13c) shows that when $C \gg K_\mu$, the rate of removal will be constant. Hence, if more Ca enters it will take longer to remove it. Moreover, the relation between the amount of Ca that enters and the time that it takes to remove it is linear. The time required to remove the entered Ca is exactly the duration of facilitation, τ_q. Hence, measurements of τ_q as a function of C_e will yield the dependence on C_e of the amount of Ca that enters per pulse, E.

Figure 9.7a, depicts a linear dependence of τ_q on C_e. However, these measurements were taken when repetitive stimulation was employed. Under such conditions, a quantitative relation between E (per pulse) and C_e is difficult to obtain. The reason is that some other factors that gain importance in the course of the repetitive stimulations might partially mask the exact dependence of E on C_e. Measurements of the dependence of τ_q on C_e when only twin pulses were employed showed a saturative dependence of τ_q on C_e. Moreover, the initial rise in τ_q at low C_e indicated that there is no cooperative relation between τ_q and C_e (Parnas *et al.*, 1982). It was therefore concluded that E shows a saturative dependence on C_e and that no cooperativity is involved. Hence E is given by (13a). For more details concerning the development of equations 13(a–c) and the experimental support for these equations, see Parnas & Segel (1989).

Final basic model for release

In our final basic model, the *release* process is governed as before by (3). *removal* will be described by a simple form of saturation with cooperativity. (It will be shown later that cooperativity in removal is not essential for predicting correctly the experimental results.) As discussed above, *entry* is described by a saturative equation without cooperativity. The following equations thus constitute our final basic model.

Entry: $E(C) = \dfrac{\varepsilon C_e}{K_\varepsilon + C_e}.$ (13a)

Release: $L(C) = \lambda \left(\dfrac{C}{K_\lambda + C}\right)^{n_\lambda}.$ (13b)

Removal: $\dfrac{dC}{dt} = -\mu \left(\dfrac{C}{K_\mu + C}\right)^{n_\mu}.$ (13c)

A numerical solution of (13a, b, c) with the parameters listed in Table 9.1 is shown in Figure 9.7(c). One can see that all the qualitative experimental results are obtained with the saturable rate of removal. By comparing the results in Figure 9.7b and 9.7c one can also conclude that the dependence of τ on C_e and hence the intersection point at $F > 1$ when decay of F is measured at different C_e, as well as the increase in F as C_e increases at longer time intervals – all these results stem from the saturation in removal and not from saturation in any other process. The reason is that the entry equation (13a) and the release equation (13b) are the same in Figures 9.7b and 9.7c, only the removal equation differs.

It can be concluded that the three equations of (13) provide a correct qualitative description of the phenomenon of facilitation and its dependence on C_e. Therefore the saturability in removal should be accepted and its consequences should be kept in mind when other aspects of facilitation are studied. The main consequence of saturability in removal is that the time course of facilitation depends on C_e, or to be more precise, on the internal Ca concentrations at the end of stimulation.

Before we move to a study of facilitation seen after a train of repetitive stimulations (PTP), two points must be made. First, as can be seen in Figure 9.9, cooperativity in removal is not essential to explain the major effects of saturability in removal. In this figure, F as a function of C_e is seen at 5 and 40 ms intervals between pulses. The results are virtually the same for $n_\mu = 1$

Table 9.1. *List of standard numerical values for the parameters*

ε	0.1 mmole l^{-1}	μ	0.0016 mmole l^{-1} ms^{-1}
K_ε	0.5 mmole l^{-1}	K_μ	0.004 mmole l^{-1} when $n_\mu = 2$,
λ	1 mmole l^{-1}		or 0.008 mmole l^{-1} when $n_\mu = 1$
K_λ	0.015 mmole l^{-1}	n_μ	1 or 2
n_λ	4 or 1	C_r	0.001 mmole l^{-1}

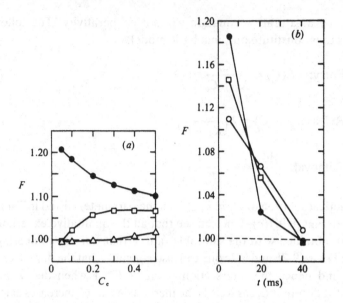

Figure 9.8. Various aspects of facilitation when no cooperativity in release is assumed. (*a*) Facilitation is a function of C_e, at time intervals of ●, 5, □, 20 and △, 40 ms. (*b*) Decay of facilitation at $C_e = $ ●, 0.1; □, 0.2 and ○, 0.4 mmole l^{-1}.

and $n_\mu = 2$; the only thing that differs is the numerical value that some of the parameters must be assigned. Another point is that cooperativity in *release* is also not important and the same basic results are obtained when $n_\lambda = 1$ as long as the removal process is a saturable one. This result can be seen in Figure 9.8 in which $n_\lambda = 1$ and we still see the intersection point and the decrease in F as C_e is raised at short time intervals, and the increase in F at longer time intervals. The exact value of n_λ determines the magntude of facilitation but not its behavior. See Parnas & Segel (1980) for a further study of the basic facilitation phenomena.

Are early and late facilitation, augmentation and post-tetanic-potentiation different processes?

One of the most important differences between the results of exponential and saturable removal is that while the first assumption predicts independence of the time course of facilitated release on the initial level of internal Ca (immediately following the stimulation), the assumption of saturability in removal predicts that the time duration increases as the initial level of internal Ca increases. Because of this difference, advocates of the exponential removal approach must explain the different time courses

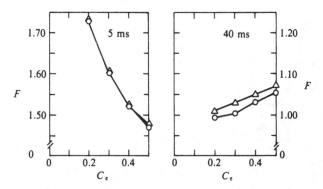

Figure 9.9. Aspects of facilitation with and without cooperativity in removal. $\triangle$, $n_\mu = 1$; $\bigcirc$, $n_\mu = 2$.

observed in different experimental conditions as caused by different processes; while advocates of saturable removal will look for different initial concentrations of internal Ca caused by the different experimental conditions.

In view of the exponential removal approach, it is not surprising that the phenomenon of increased transmitter release following a test pulse has been given at least four names, each corresponding to a different process which is believed to take place. Figure 9.10 summarizes the situation.

One can see that the durations of facilitated release differ by orders of magnitude when the experimental conditions differ. Thus when one pulse is given before the test pulse, C_e is 0.2 or 0.5 mmole l^{-1}, and when the external Mg level, M_e, is 1 mmole l^{-1}, then the duration of facilitation is 30–50 ms. If a *few* preliminary pulses are given, if C_e is 1.8 mmole l^{-1} and if M_e is greatly increased to 12–17 mmole l^{-1}, the duration is already several hundreds of ms. Because these observed phenomena could not be described by a single exponent, Mallart & Martin (1967) proposed that **early** and **late facilitation** were two processes occurring at different times, with total facilitation being the sum of the two individual facilitations.

When a long train of repetitive stimulation is given, the duration is seconds, or even minutes in other experiments, and here too the whole period of decay cannot be fitted by one exponent. As a result, two more processes have been suggested. **Augmentation** follows the repetitive stimulation and has a time course of several seconds, while **PTP** happens later with a time course of tens of seconds or even minutes.

It should be emphasized that the 'different processes' postulated by the various authors do not necessarily require that a factor different from internal Ca is involved, but that at different times different processes

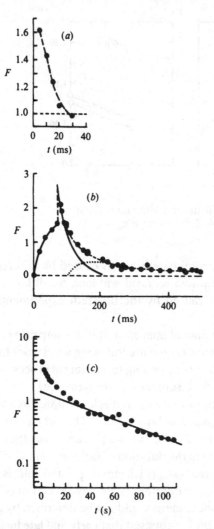

Figure 9.10. Typical (exponential) early and late facilitation (as well as augmentation) and potentiation. (*a*) Facilitation with a time course of 30–50 ms (from Rahamimoff, 1968). (*b*) Early and late facilitation (from Mallart & Martin, 1967). (*c*) Augmentation and potentiation (Magleby & Zengel, 1976).

regulate the internal Ca concentration (Erulkar & Rahamimoff, 1978) or that more than one pool of Ca is involved (Magleby, 1973). According to the common view, then, we have at least four processes that contribute at different times to facilitated release, and the decay of facilitation represents the sum of those processes.

Is saturation in removal enough to unify facilitation and PTP?

If we are to adhere to the saturation model we must try to explain the long time course of PTP with the equations of (13). Moreover it is seen experimentally (Figure 9.10) that both the decay of facilitation after one pulse and the decay of PTP sometimes has two time constants. Could saturation in removal account for this? We shall now examine this question.

According to the rules so far, PTP and facilitation both depend on the same pool of free internal Ca and the decay of both depends on the initial level of internal Ca and the rate of its removal. In the case of repetitive stimulation, the initial internal Ca concentration is normally higher than after one pulse, therefore a longer time course for PTP is expected. However, if the initial level of internal Ca is set equal to the external concentration (in the frog by using Ringer with $C_e = 1.8$ mmole l^{-1}) one can calculate the longest possible duration τ that can be expected with the given parameters, on the basis of saturation in removal only. The maximum τ is found to be around 1 s, which is much too short to account for PTP.

Another shortcoming of the present theory is revealed by simulating a train of five pulses with the equations of (13). Figure 9.11a, shows the experimentally observed increase in facilitation during a train of five pulses and its decay after stimulation (Mallart & Martin, 1967). There are three basic differences.

(a) The computed increase in F is too small. This discrepancy is even larger than it appears since Mallart & Martin (1967) define F as $(L_2/L_1) - 1$.

(b) The initial decay rate is too slow in the computed results.

(c) The duration of facilitation is too short.

It is apparent that saturation in removal, even though it does explain correctly all types of dependence of facilitation on Ca concentration, is not enough to account for some of the results seen after a train of stimulations. Examination of the three differences between the computed and experimental results shown in Figure 9.11 reveals that both the low F which is attained during stimulation and its lower decay immediately following stimulation might result from the same source, namely higher $C(0)$ in the computations than in the experiment. One can see from Figure 9.7 that when $C_e = 0.2$ mmole l^{-1}, both F and its initial decay rate were higher than when $C_e = 0.4$ mmole l^{-1}. This means that in the calculations, too much Ca was allowed to enter.

Why, then, is τ too short? It may be that the calculated rate of removal is too fast, but what can account for this? If one compares the experimental conditions of Rahamimoff (1968), when one pulse is given, to those of

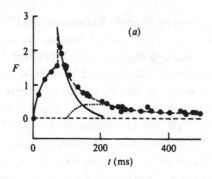

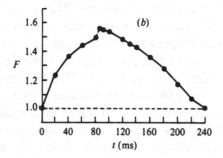

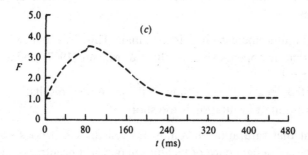

Figure 9.11. Experimental and calculated build up and decay of
facilitation. (*a*) Experimental build up and decay of facilitation
(Mallart & Martin, 1967). (*b*) Build up and decay of facilitation
calculated with the equations of (14) and the set of parameters
listed in Table 9.1. (*c*) Build up and decay of facilitation
calculated with (14)–(16), parameters: $\bar{K}_\varepsilon = 2$ mmole l^{-1},
$C_e = 1.8$ mmole l^{-1}.

Mallart & Martin (1967), one can see that there is a major difference. While
in Rahamimoff's experiments the external concentration of Mg was
1 mmole l^{-1}, it was 12–17 mmole l^{-1} in the five pulse experiments of Mallart
& Martin (1967).

As was mentioned earlier, the role of Mg is to block transmitter release. When a train of repetitive stimulation is given, transmitter may be released so frequently and in such large amounts that the internal pools of transmitter become exhausted. In such an event, the measured facilitation would be an artifact of the depletion of internal stores of transmitter. In addition to this, if a nerve–muscle preparation is used then high release can cause a strong contraction of the muscle and thus a jump of the microelectrode used to measure the e.p.p. High external concentrations of Mg are therefore employed in many of the experiments in which repetitive stimulation is given. (It should be mentioned that Mg is not the only means used to overcome the two problems just referred to, but it is a very common one.)

Although the ability of Mg to reduce the amount of transmitter released has often been exploited, its mode of action has not been investigated until recently. In order to inhibit release, if release indeed reflects the internal Ca concentration, a substance must either inhibit the entry of Ca into the nerve terminal during stimulation, or inhibit release, or both. If Mg does inhibit release, it should itself enter the nerve terminal. If Mg competes with Ca on entry, this will explain why the calculated concentration of Ca that entered was too high (in the 5-pulse experiments) and thus F was too low and its initial decay too slow. The shorter calculated duration cannot be explained by that competition.

It has been found that Mg indeed inhibits the entry of Ca (Baker, 1975). Furthermore Rojas & Taylor (1975) showed that Mg does enter the nerve terminals together with Ca. According to their calculations, more Mg enters than Ca. Miledi (1973) injected Mg together with Ca into squid axon. He compared release with injected Ca alone to that with Ca and Mg together, and found that release was slightly inhibited in the presence of Mg. The main inhibition of release caused by Mg seems thus to be the result of competition on entry of Ca. Carafoli & Crompton (1975) showed that Mg competes with Ca on uptake into heart mitochondria. On the basis of these facts, it is suggested here that Mg competes with Ca in each of the three processes: entry, release and removal.

The competition of Ca and Mg on removal is an extension of the saturation in removal rate. We suggest that this competition results in the long durations of facilitation that occur after a train of stimulation, during which the internal concentration of both Ca and Mg are raised. It is not the intention here to say that only Mg is able to cause long durations of facilitation. It is suggested that any causes of saturation in removal together with a decrease in the rate of removal can result in prolonging the duration τ by orders of magnitude.

There are other ways to reduce the rate of removal such as lowering the temperature, or by competition between Ca and other ions that are known to enter during nerve stimulation. The result will be the same – prolongation of τ. Mg is chosen for further theoretical investigation since it is such a common component of the experiments in which a long train of stimulations is given. We shall use $M(t)$ to denote the internal concentration of Mg, and M_e to denote the external concentration. An overbar will be employed to distinguish the various constants and functions pertaining to Mg from the corresponding constants and functions for Ca.

Competition between Ca and Mg on entry, release, and removal

We shall take into account the competition between Ca and Mg by generalizing the equations of (13) as follows:

$$\text{Entry of Ca:} \quad E(C, M) = \frac{\varepsilon}{\dfrac{K_\varepsilon}{C_e}\left(1 + \dfrac{M_e}{\bar{K}_\varepsilon}\right) + 1}. \tag{14a}$$

$$\text{Entry of Mg:} \quad \bar{E}(C, M) = \frac{\varepsilon}{\dfrac{\bar{K}_\varepsilon}{M_e}\left(1 + \dfrac{C_e}{K_\varepsilon}\right) + 1}. \tag{14b}$$

$$\text{Release:} \quad L(C, M) = \frac{\lambda}{\left[\dfrac{K_\lambda}{C}\left(1 + \dfrac{M}{\bar{K}_\lambda}\right) + 1\right]^{n_\lambda}}. \tag{15}$$

Mg is not effective in release (Miledi, 1973), but it can bind to the release sites with a dissociation constant $\bar{K}_\lambda$.

$$\text{Removal of internal Ca:} \quad \frac{dC}{dt} = \frac{-\mu}{\left[\dfrac{K_\mu}{C}\left(1 + \dfrac{M}{\bar{K}_\mu}\right) + 1\right]^{n_\mu}}. \tag{16a}$$

$$\text{Removal of internal Mg:} \quad \frac{dM}{dt} = \frac{-\bar{\mu}}{\left[\dfrac{\bar{K}_\mu}{M}\left(1 + \dfrac{C}{K_\mu}\right) + 1\right]^{n_\mu}}. \tag{16b}$$

For simplicity, the degree of cooperativity has been assumed to be the same for Ca and Mg. Such an assumption seems justified in view of our past experience that the details of the equations have little effect.

If we reexamine the previously found defects of the model, we see that they are abolished by introducing competition in removal in addition to saturation. The duration τ increases by orders of magnitude as both $C(0)$ and

Table 9.2. *Durations of facilitated release (τ) under various conditions**

C_e mmole l^{-1}	M_e mmole l^{-1}	No. of impulses	$C(0)$ mmole l^{-1}	$M(0)$ mmole l^{-1}	τ ms
0.2	1	1	0.028	1	116
0.5	1	1	0.049	1	137
0.2	1	50	0.177	1	237
0.2	20	50	0.15	19.9	5600
1.8	20	10	0.148	4.9	724

* See text for discussion of parameters

$M(0)$ increase, as can be seen in Table 9.2. The effect of saturation is seen if one compares results when $C_e = 0.2$ mmole l^{-1} for 1 and 50 pulses; τ increased from 116 to 237 ms. A more striking change is caused by the competition in removal. If one compares $C_e = 0.2$ mmole l^{-1} in the presence of 1 and 20 mmole l^{-1} of external Mg, when 50 pulses are given τ changes from 237 to 5600 ms. The numerical values of the parameters have not yet been studied so that it is not possible at present to give a fully meaningful comparison between these and the previously described results (Ca alone).

At the moment we are content to show that τ can change by orders of magnitude if the experimental conditions are such that both Ca and its competitor on removal can accumulate internally. Such conditions exist in repetitive stimulation. In Figure 9.12, a decay of facilitated release is seen

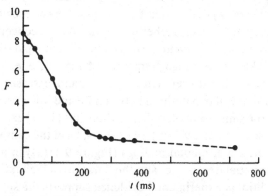

Figure 9.12. Decay of facilitated release following a train of 10 impulses with high M_e. Calculated with (14)–(16) and parameters as in Figure 9.11c, except that $M_e = 20$ mmole l^{-1}.

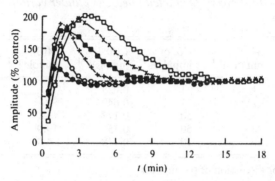

Figure 9.13. Decay curves of PTP following tetani of different frequencies (Rosenthal, 1969). ●, 40 s^{-1}, ○, 50 s^{-1}; +, 62.5 s^{-1}, ■, 77 s^{-1}; ×, 100 s^{-1}; □, 125 s^{-1}.

which corresponds to $C_e = 1.8$ mmole l^{-1} and $M_e = 20$ mmole l^{-1} when 10 pulses are given. Two distinct decay rates, i.e. two 'time constants' are clearly seen, with the first much larger than the second. Because of the high value of M_e and the relative small number of pulses, $C(0)$ is relatively low while $M(0)$ is relatively high, which results in a decay curve with two time constants. This result shows that *more than one time constant in a decay curve does not necessarily mean that the sequential appearance of more than one process is responsible for the decay. With appropriate parameter choices, competition on removal can result in the same shaped curve.*

Further support for the idea that the shape of the decay curve reflects $C(0)$ and the initial rate of removal, can be obtained from the experimental results of Rosenthal (1969). In Figure 9.13, we present decay curves of PTP following tetani (pulse trains) of different frequencies. One can see that the higher the frequency, which means higher initial internal Ca concentration $C(0)$, the slower is the initial decay and the less distinct is the separation into two time constants. At the very high frequency of 125 s^{-1}, one can even observe a decay curve which is slower in the beginning and increases later on when C is reduced. Finally, if competition between Ca and Mg is used in the three processes and the simulation of Mallart & Martin's (1967) experiment is repeated, one can see in Figure 9.11(c) that the results of the computation fit reasonably with the experimental findings (Figure 9.11(a)) in all three aspects: size of F, its initial decay rate, and the duration of facilitation.

The following additional experimental evidence supports the suggestion made here that all results derive from a single set of processes.

(i) As was shown in Figure 9.7, both F and PTP show an increase in τ as

C_e is raised. If PTP results even in part from a process that does not connect internal to external Ca, such similar dependence would be difficult to explain.

(ii) The duration of both F and PTP have the same temperature dependence. Their Q_{10}'s have the same value, namely around $Q_{10} \approx 4$ (Eccles *et al.*, 1941; Magleby & Zengel, 1976).

Preliminary simulations of (14)–(16) indicate that a set of parameters that show good agreement with two-pulse experiments will enable a τ of seconds following a train of repetitive stimulation, but not minutes. This difficulty is expected since the model presented here is incomplete. The postulated removal process is a lumping of several processes. The reduction in the rate of removal that is suggested here is only because of an increase in the K_m of uptake into the mitochondria, caused by accumulation of internal Mg.

It was shown that external Mg slows the extrusion of C by means of the Na–Ca exchanger (I. Parnas *et al.*, 1982). External Mg (as well as high external Ca) blocks the exchanger. Under high concentrations of M_e, facilitation was prolonged from about 100 ms to about 1 s (I. Parnas *et al.*, 1982). However, this prolongation is independent of the number of pulses, while the reduction caused by internal accumulation depends on this number. It is also known that Na enters during stimulation and it has been suggested that internal Na slows down the uptake of internal Ca into the mitochondria. Under normal conditions, external Na is present and accumulates in the nerve terminals during repetitive stimulation. Its internal accumulation causes reduction in the rate of removal of internal Ca. Therefore the rate of removal of internal Ca, following tetanic stimulation is even lower than expected on the basis of inhibition by internal Mg only. There is thus no intention to say that only internal Mg causes PTP, and that (14)–(16) fully describe facilitation and PTP. But the theory does unify all the suggested processes involved in facilitated release into one common process.

The conclusions are as follows:

(a) In many cases in which high external Mg was present and repetitive stimulation was given, the accumulation of internal Mg during the stimulation and its competition on removal are the causes for the very long durations of facilitation.

(b) The initial concentration of internal Ca together with its *rate of removal* explains the whole period of facilitated release.

(c) Early and late facilitation as well as augmentation and PTP do not necessarily differ from each other. They may just represent different experimental conditions, and, because of this, different initial concentrations and rates of removal. Both higher external concentrations (of Ca and Mg) and (more importantly) repetitive stimulation increase the initial concentrations of internal Ca and Mg, and thus prolong the duration of F and also change its decay shape.

References

Alnaes, E. & Rahamimoff, R. (1975). On the role of mitochondria in transmitter release from motor nerve terminals. *J. Physiol.* **248**, 285–306.

Baker, P. F. (1975). The regulation of intracellular calcium. *Soc. Exp. Biol. Sym.* **30**, 67–88.

Balnave, R. J. & Gage, P. W. (1974). On facilitation of transmitter release at the toad neuromuscular junction. *J. Physiol.* **239**, 657–75.

Balnave, R. J. & Gage, P. W. (1977). Facilitation of transmitter secretion from toad motor nerve terminals during brief trains of action potentials. *J. Physiol.* **266**, 435–51.

Blaustein, M. P. & Hodgkin, A. L. (1969). The effect of cyanide on the efflux of calcium from squid axons. *J. Physiol.* **200**, 497–527.

Carafoli, E. & Crompton, M. (1975). Calcium ions and mitochondria. *Soc. Exp. Biol. Sym.* **30**, 89–114.

Connor, J. A., Kretz, R. & Shapiro, E. (1986). Calcium levels measured in a presynaptic neuron of Aplysia under conditions that modulate transmitter release. *J. Physiol.* (London) **375**, 625–42.

Del Castillo, J. & Engbaek, L. (1954). The nature of the neuromuscular block produced by magnesium. *J. Physiol.* **124**, 370–84.

Del Castillo, J. & Katz, B. (1954a). Quantal components of the end-plate potential. *J. Physiol.* **124**, 560–73.

Del Castillo, J. & Katz, B. (1954b). Statistical factors involved in neuromuscular facilitation and depression. *J. Physiol.* **124**, 574–85.

Dipolo, R. (1973). Calcium efflux from internally dialyzed squid giant axons. *J. Gen. Physiol.* **62**, 575–89.

Dodge, F. A., Jr & Rahamimoff, R. (1967). Cooperative action of calcium ions in transmitter release at the neuromuscular junction. *J. physiol.* **193**, 419–32.

Eccles, J. C., Katz, B. & Kuffler, S. W. (1941). Nature of the 'end plate potential' in curarized muscle. *J. Neurophysiol.* **4**, 362–87.

Erulkar, S. D. & Rahamimoff, R. (1978). The role of calcium ions in tetanic and post-tetanic increase of miniature end-plate potential frequency. *J. Physiol.* **278**, 501–11.

Katz, B. (1965). *Nerve, Muscle and Synapse*, New York, McGraw-Hill Publishing Company.

Katz, B. & Miledi, R. (1967a). A study of synaptic transmission in the absence of nerve impulses. *J. Physiol.* **192**, 407–36.

Katz, B. & Miledi, R. (1967b). The timing of calcium action during neuromuscular transmission. *J. Physiol.* **189**, 535–44.

Kuffler, S. W. & Nicholls, J. G. (1976). *From Neuron to Brain*, Sunderland, Mass., Sinauer Assoc., Inc.

Magleby, K. L. (1973). The effect of tetanic and post-tetanic potentiation on facilitation of transmitter release at the frog neuromuscular junction. *J. Physiol.* **234**, 353–71.

Magleby, K. L. & Zengel, J. E. (1976). Augmentation: a process that acts to increase transmitter release at the frog neuromuscular junction. *J. Physiol.* **257**, 449–70.

Mallart, A. & Martin, A. R. (1967). An analysis of facilitation of transmitter release at the neuromuscular junction of the frog. *J. Physiol.* **193**, 679–94.

Miledi, R. (1973). Transmitter release induced by injection of calcium ions into nerve terminals. *Proc. R. Soc. Lond. B* **183**, 421–85.

Miledi, R. & Thies, R. E. (1967). Post-tetanic increase in frequency of miniature end-plate potentials in calcium-free solutions. *J. Physiol.* **192**, 54–55P.

Parnas, H., Dudel, J. & Parnas, I. (1982). Neurotransmitter release and its facilitation in crayfish. I. Saturation kinetics of release, and of entry and removal of calcium. *Pflügers Arch.* **393**, 1–14.

Parnas, H. & Segel, L. A. (1980). A theoretical explanation for some effects of calcium on the facilitation of neurotransmitter release. *J. Theoret. Biol.* **84**, 3–29.

Parnas, I., Parnas, H. & Dudel, J. (1982). Neurotransmitter release and its facilitation in crayfish. II. Duration of facilitation and removal processes of calcium from the terminal. *Pflügers Arch.* **393**, 232–6.

Parnas, H. & Segel, L. (1989). Facilitation as a tool to study the entry of calcium and the mechanism of neurotransmitter release. *Progress in Neurobiology* **2**, 1–9.

Parnas, H., Parnas, I. & Segel, L. A. (1990). On the contribution of mathematical models to the understanding of neurotransmitter release. *Int. Rev. Neurobiol.* (in press).

Rahamimoff, R. (1968). A dual effect of calcium ions on neuromuscular facilitation. *J. Physiol.* **195**, 471–80.

Rojas, E. & Taylor, R. E. (1975). Simultaneous mesurements of magnesium, calcium and sodium influxes in perfused squid giant axons under membrane potential control. *J. Physiol.* **252**, 1–27.

Rosenthal, J. (1969). Post-tetanic potentiation at the neuromuscular junction of the frog. *J. Physiol.* **203**, 121–33.

Silinsky, E. M., Mellow, A. M. & Phillips, T. E. (1977). Conventional calcium channel mediates asynchronous acetylcholine release by motor nerve impulses. *Nature, Lond.* **270**, 528–30.

10

Acceptable and unacceptable models of liver regeneration in the rat

Summary

If large portions of the rat liver are removed, the remainder shows a surprising capacity to regenerate and this chapter discusses the two most likely mechanisms by which this could happen. The first assumes that there is a tissue-specific growth inhibitor (a chalone) of short half life that is normally produced by the liver and whose concentration inevitably drops after partial extirpation, so allowing cells to divide and the liver to increase in size and produce more inhibitor. The second assumes that some waste product which the liver normally destroys can also act as a growth stimulator. The build-up of this compound in the presence of a functionally inadequate amount of liver will thus cause the remaining liver cells to divide until the liver is again big enough to degrade all the waste product as it is made. Differential equations describing the two models are derived and the forms of their solution are examined analytically and numerically. It turns out that only the first model is able to describe the observed kinetics of regeneration as the second predicts that the regenerating liver should overshoot its target size.

Introduction

The cells of the rat liver rarely divide in the normal adult. If, however, in an operation called a partial **hepatectomy**, ~65% of the liver is removed by surgical ablation of the two larger lobes, there is a great burst of mitosis in the cells of the remaining two lobes that peaks some 26 h after the operation and then declines. Within two weeks or so, these lobes grow to the size of a normal liver and the system becomes quiescent (see Bucher, 1963, and Alison, 1986, for review). This phenomenon has been much studied but the mechanism responsible for regeneration has not yet been elucidated.

There are, however, two plausible mechanisms that, it is thought, might work. First, the liver could continually make an inhibitor of cell division (see e.g. Glinos, 1958), the concentration of which would normally be high enough to suppress mitosis. Following hepatectomy, the concentration of this inhibitor in the blood and liver remnant would drop as the molecule is assumed to have a short half-life. As a result of this decrease, mitosis would occur, the liver would grow, more inhibitor would be made and the *status ante quo* would be restored.

The second theory suggests that liver size is controlled by function rather than by size so that, after hepatectomy, the functionally inadequate liver grows until it can fulfill all its biochemical tasks (see e.g. Goss, 1964). The mechanism by which this is thought to be achieved is that some waste product builds up in the presence of an inadequately sized liver and this stimulates mitosis; as the liver increases in size so the waste product level drops and, eventually, the normal liver size is restored.

In this chapter, simple mathematical models that incorporate these two mechanisms are put forward and the extent to which each can explain the known experimental facts is examined. In particular an attempt is made to find functional parameters of the inhibitor and stimulator that allow the best match of the prediction of each model to the published data on regeneration. For a rather more detailed analysis of liver regeneration, see Bard, 1978 and 1979.

Experimental data

A great deal of work has been published on liver regeneration and this section mentions only the facts essential for the analysis. Observations have been made in four contexts which to some extent overlap.These are studies on the rates and localization of mitosis in regenerating liver, physiological experiments designed to see how regeneration is controlled, the isolation of stimulators or inhibitors of mitosis in the liver and attempts to interfere with regeneration by giving drugs or other compounds to animals. The results of the last class may be briefly summarized: chemicals have been found that mimic to a limited extent the effects of regeneration in the intact liver (e.g. α-hexachlorocyclohexane, see Schulte-Hermann, 1977), but their mode of action is unknown.

The most important data on regeneration are those that describe the rates at which cells divide in the regrowing liver. As a typical set of observations here, we can consider the work of Grisham (1962), who measured mitotic indices every few hours for the three days after 65% hepatectomy. He found that after an 18 h lag the mitotic index rose and then slowly dropped (Figure 10.1). In the accompanying autoradiographical study he showed that it was the parenchymal cells (~90% of the liver mass) that initially took up [^{3}H]thymidine and that the other cells of the liver divided later. More recently, Fabrikant (1968) has shown that the parenchymal cells in a different breed of rats show a second peak of mitosis at around 56 h after surgery.

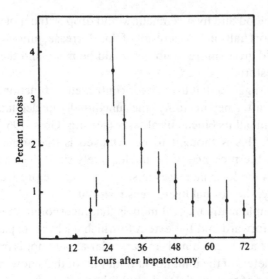

Figure 10.1. Mitotic indices in the regenerating livers of rats after 65% hepactectomy as measured by Grisham (1962) (from Bard, 1978).

The mitotic response when small amounts of liver are removed is too small to measure accurately and growth has to be assayed indirectly by the uptake of radioactive thymidine. The major new observation made with this technique (see e.g. Bucher, 1963) is that the peak of the uptake rises steeply as increasing amounts of liver are removed (Figure 10.2). There is a further interesting observation that has been made from autoradiographic studies using [^{3}H]thymidine uptake: Fabrikant (1968) and Rabes, Wirshing, Tuczek & Iseler (1976) have shown that cells near the periportal areas of the liver, where the blood enters, divide sooner than those near the hepatic veins, where blood drains away from the parenchymal cells.

Physiological studies have now shown, after considerable controversy, that there are components in the blood that play a role in regeneration. Sakai (1970) showed that, in parabiotic rats (i.e. pairs of animals with their blood supplies linked), where one partner has had a large partial hepatectomy, the liver of the control rat takes up considerable amounts of [^{3}H]thymidine. This experiment proves that there are blood-borne factors in regeneration but not whether the response is due to inhibitor loss or stimulator gain. Some early, preliminary experiments by Glinos (1958) indicate that it is inhibitor loss. He replaced part of a rat's plasma with saline in order to dilute any factor present: he observed a burst of mitosis, a result incompatible with a stimulator of mitosis being present.

As to the nature of the components in the blood that mediate liver regeneration, there is now a considerable body of evidence to show that one at least is an inhibitor of hepatic mitosis (for review, see Alison, 1986). Such tissue-specific mitotic inhibitors are known as chalones (Bullough, 1962) and perhaps the most interesting evidence for such a molecule operating in the G_1 phase of the liver-cell cycle comes from the work of McMahon, Farrelly & Iype (1982). They isolated a protein from normal rat livers which inhibits reversibly the growth of normal but not malignant liver cells *in vitro*. This molecule has a molecular weight of about 26 000 and is effective at concentrations of 1–10 nM. Such an inhibitor is an excellent candidate for mediating liver regeneration; unfortunately, no one has yet shown that, when a liver is partially extirpated, the concentration of this molecule in the blood drops.

There is one other piece of information that is worth mentioning in this context because it allows us to discuss briefly a third model of regeneration. At first sight, the removal of large amounts of liver might cause a *mitotic stimulator* to be generated with this wound hormone underpinning regeneration, but a moment's thought shows that this mechanism alone is implausible. As we have already noted, Bucher (1963) has shown that the more liver is removed, the greater is the mitotic response of the remainder. Were regeneration to be under the influence of a stimulator, there would have to

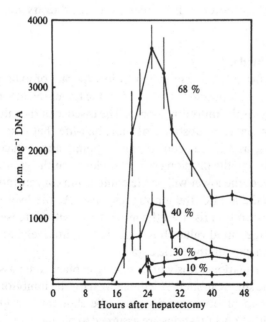

Figure 10.2. [³H]Thymidine incorporation rates in the livers of young rats after various amounts of hepatectomy (from Bard, 1978). The data are taken from the experiments of Bucher (1963).

be some mechanism to increase stimulator production disproportionately with the amount of liver removed, and it is very difficult to envisage a convincing mechanism by which this could be achieved. Nevertheless, LaBrecque *et al.* (1978) have isolated a compound from the intact livers of young rats which stimulates liver growth *in vivo* and *in vitro*. At the moment, the status of this molecule is unclear, but, as it is found in young, rapidly growing rats and as its concentration in the liver shows a diurnal rhythmicity, the function of this compound may be to facilitate normal rather than compensatory growth.

A note on units

Below, equations will be derived for the two models which describe how the liver could grow as the concentrations of inhibitor or stimulator vary. To simplify calculations of concentration as the liver size changes, we will consider the volume v of the liver rather than its mass. For a 150 g rat, the liver weighs about 6 g (Grisham, 1962) and thus has a volume of 6 ml; the blood volume is 15 ml. In such a liver, the resting mitotic index is about 0.0004% (i.e. in sections of normal liver, some 4 cells in 10^4 are seen to be in mitosis) and mitosis itself takes about 1 h (Fabrikant, 1968). It is therefore convenient to measure time in hours, and dv/dt is thus the mitotic rate per milliliter of liver. Concentration will be measured in arbitrary units per milliliter.

The inhibitor model

The essential facts that any model must incorporate or explain are that there is a blood-borne humoral factor and that the larger the amount of liver removed the larger is the mitotic response. The essence of the inhibitor model is that the liver makes a substance of short half-life that is secreted into the blood (this is, as will be seen, a crucial point) and reduces the likelihood of a cell entering the division cycle. It is clear that the greater the amount of liver removed, the lower will be the equilibrium concentration of the inhibitor and thus the greater the mitotic response. As the liver grows and the inhibitor concentration rises, so the mitotic rate will decrease from its peak and return to the initial value. It is clear that qualitatively, at least, the inhibitor model can explain the basic data.

Before deriving the equations, it is worth making explicit some assumptions that lie behind the simple model. The time taken for the inhibitor to be transferred from liver to blood is considered to be short and, a greater assumption, the inhibitor concentrations are assumed to be the same in both liver and blood. The inhibitor itself exerts an immediate concentration–dependent effect on a cell so that the likelihood of a cell entering the mitotic cycle can be defined by a simple dose–response function. Such cells are assumed to start enlarging immediately after the decision is made [there is evidence to support this (Tongendorff, Trebin & Ruhenstroth-Bauer, 1975)] so that dv/dt represents both the immediate increase in size and the mitotic rate at the end of the cell cycle (assumed constant for all cells). Finally, it is necessary to assume that the mitotic rate never gets so high that a cell would have to divide a second time before it had completed a first mitosis; this point is considered in more detail a little later.

For the model, let the liver be homogeneous having a volume v and make an inhibitor at a rate p per unit volume the name and concentration of which are c and whose decay constant is q (so that the half-life is $0.69/q$). The inhibitor is secreted into the blood (volume w) and the equilibrium values of inhibitor and liver size are c_0 and v_0. The rate of liver growth is given by an unknown function $f(c)$ and the rate of cell death in the liver growth is given by r (assumed to equal the resting mitotic rate). The role of r in these equations is to define the resting equilibrium. The rate of growth is thus given by

$$dv/dt = v[f(c) - r],\qquad(1)$$

and the mitotic index is $(1/v)\,dv/dt$. As inhibitor is distributed uniformly between liver and blood, the rate at which its concentration grows is given by

$$dc/dt = [pv/(v + w)] - qc.\qquad(2)$$

It is worth noting the role of the blood volume in these equations: if $w = 0$, then the inhibitor concentration is independent of liver size and there can be no regulation. The blood thus acts as a reservoir of inhibitor.

The condition that these equations give a stable equilibrium is that $f(c)$ be an inhibitor so that $f'(c) < 0$. It turns out that there is a direct return to this point under physiological conditions. Using the optimal value of $f(c)$ which will be determined later, the condition that the equilibrium point is a node rather than a focus turns out to be that $q > 4c_0r$ or that the half-life of the inhibitor is greater than 100 h or so.

Inserting those parameters that can be measured into the equations, they become

$$dv/dt = v[f(c) - 0.0004],\qquad(3)$$

$$dc/dt = [pv/(15 + v)] - qc,\qquad(4)$$

and it can be seen that if $f(c)$ and q, the functional parameters of the inhibitor, are inserted, then the value of p can be determined from the equilibrium conditions. We must therefore find values of $f(c)$ and q which, when substituted, will give solutions that match the data. The time between entry into cell cycle and mitosis must also be found so that the delay between hepatectomy and mitosis can be allowed for. If all this can be done, the model is shown to be at least plausible; if parameters cannot be found then the model must be rejected.

The appropriate form for the dose–response function is not immediately clear, but some constraints on it come from the data shown in Figures 10.2 and 10.5. They show first that, as inhibitor concentration drops to about one third of its value, the mitotic rate rises by two orders of magnitude, and secondly that the response is nonlinear. Various dose–response functions have been tried and the only simple one that gives solutions that match the data is, as might be expected from the data:

$$f(c) = A \exp(-\alpha c),$$

where for convenience we let $\alpha = 1$ and have dimension $[\text{conc}]^{-1}$

A computer program has been written to solve (3) and (4) when values of c_0 and q are given. Trial and error show that the optimal match of theory to data is given when $c_0 = 8.8 \pm 0.2$ and the decay constant is $0.25 \pm 0.03\,\text{h}^{-1}$ (implying a half-life of $2.9 \pm 0.4\,\text{h}$) and the time between entering the mitotic cycle and cell division has to be 15 h (a reasonable time for the cell cycle). If the decay constant is greater, the rise in mitosis is faster, whereas with a lower decay constant the mitotic rate rises more slowly. The height of the peak is determined by $f(c)$ if the half-life is kept constant.

The match between theory and data is quite good when the optimal parameters are substituted in the equations: the mitotic index rises rapidly to a peak and then declines slowly (Figure 10.3). Quantitatively, the predicted peak is not as sharp as that in the data and the rate of descent rather greater. The reason for this is that $f(c)$ acts on all cells in the liver and includes cells that have already entered the cell cycle. However, given the simplicity of the model, the fit is acceptable. Over several days (Figure 10.4), the liver size increases rapidly and returns to 85% of its original size in about ten days. The inhibitor concentration likewise increases after its initial decay.

The model can be improved to overcome the major inadequacy of the analysis, the unjustified assumption that all the cells in the regenerating liver are available to enter the mitotic cycle at any time. This is clearly not so because, once a cell has entered the mitotic cycle, it takes about 15 h to traverse it and divide into two cells; only after these cells have been through a rest period, G_0, can they re-enter the cycle. In analysing the kinetics of regenerating liver, we therefore have to ensure that only those cells that have not entered the mitotic cycle are subject to the effects of the inhibitor. This factor can be built into the equations with some trouble but little

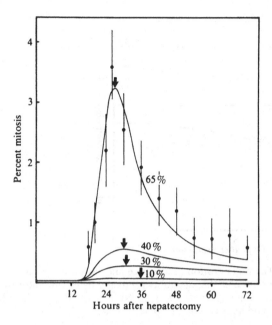

Figure 10.3. Theoretical mitotic indices predicted by the inhibitor theory, using a half-life of 2.9 h, a negative exponential dose–response curve and a 15 h delay between entry to the mitotic cycle and mitosis (from Bard, 1978). The data points of Grisham (1962) are matched quite well; the form of the curves is similar to the data on [³H]thymidine incorporation obtained by Bucher (1963). The arrows show the times of maximal mitosis.

difficulty (for details, see Bard, 1979) and, if best-fit parameters are re-chosen, the predicted kinetics match those observed quite well (Figure 10.5). The most noticeable improvement in the analysis is that the solutions now predict the second small burst of mitosis that is usually observed, while the major change in the parameters required to effect this improvement is the lengthening of the inhibitor half life from 3 h to 11.4 h.

The mitotic rate prediction of the theory for ablations of less than 65% cannot be matched directly as, for such operations, the only data are on thymidine incorporation studies. However, if it is assumed that such incorporation is proportional to the mitotic rate, then a scaled comparison between the predicted mitotic index and the incorporation rates can be made (Figure 10.6). The theoretical curve matches well the data of Bucher (1963) and shows the threshold effect that is expected. The form of the curve derives, of course, from the nonlinear dose–response curve.

In summary, therefore, the inhibitor model can explain the data on liver

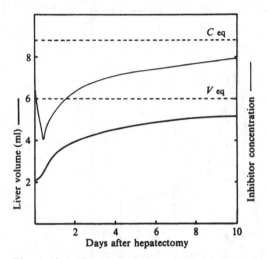

Figure 10.4. The regeneration of liver size and restoration of inhibitor concentration after 65% hepatectomy predicted by the inhibitor theory with optimal parameters (from Bard, 1978). C_{eq} and V_{eq} are the inhibitor concentration and liver volume, respectively, at equilibrium.

regeneration provided that the inhibitor has a half-life of a few hours and a negative exponential dose–response curve.

Function-dependent regeneration

In this approach, it is assumed that liver growth after partial extirpation is stimulated by a waste product that would be almost completely degraded by the intact liver (Goss, 1964). The greater the concentration of waste product the greater will be the stimulus for liver cells to divide. Again, this model appears to describe the essential features of regeneration. However, a deeper analysis shows that there are some surprising aspects to this mechanism.

Again, let the liver volume be v and the blood volume be w; let a waste product c' be generated by the body at a rate s and let it be degraded by the liver at a rate u per unit volume, the waste product having the same concentration in both liver and blood. The assumptions that were discussed for the inhibitor case are assumed to hold here. If the rate at which the liver cells divide is given by the function $g(c')$, which increases with c' so that $f'(c') > 0$ then, as before,

$$dv/dt = v(g(c') - r), \tag{5}$$

and the rate at which the waste product builds up is given by

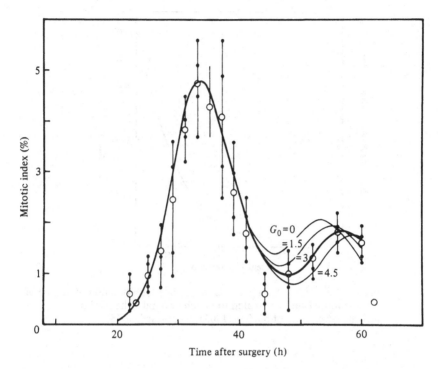

Figure 10.5. Matching the solutions of the 'inhibitor' model incorporating cell-cycle parameters to the observations of Fabrikant (1968) on regenerating rat liver in August strain rats. The small dots are data points and the circles are mean values. The thick line shows the optimal fit of the model and assumes that the inhibitor half-life is 11.4 h, the cell-cycle time is 18.25 h and the resting period G_0 is 3 h. The fine-line solutions show that the effect of varying G_0 is to alter the timing of the second burst of mitosis (from Bard, 1979).

$$dc'/dt = \frac{s - uv}{v + w} = \frac{u(v_0 - v)}{v + w}, \tag{6}$$

as $s = uv_0$ at equilibrium.

Although these equations have an equilibrium point at some (v_0, c_0'), this point is not stable as can be seen by linearizing these equations around (v_0, c_0'). For this, let $v = v_0 + x$ and $c' = c_0' + y$; then

$$dx/dt = 0 + v_0 g'(c')y \quad \text{[as } g(c_0') = r \text{ by definition]},$$

$$dy/dt = [-ux/(v_0 + w)] + 0.$$

Thus it can be seen that the equilibrium point is a center rather than a node

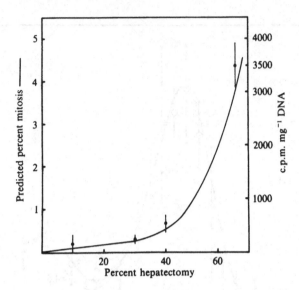

Figure 10.6. Theoretical peak mitotic indices as a function of degree of hepatectomy (line) are scaled to match optimally the [³H]thymidine incorporation data of bucher (1963) (from Bard, 1978).

or focus, and the trajectories of the system in (c, v)-space are a series of ellipses around this point. Qualitatively, this means that, if part of the liver is removed, the remainder, instead of growing back to its original size, overshoots and becomes larger than it was in order to deal with the backlog of waste product that had accumulated while the liver was undersized. The now oversized liver will reduce the amount of waste product to below its original concentration so that mitosis in the liver is below the death rate and the liver eventually becomes smaller than the equilibrium size; the waste product will now build up and the cycle will start again.

The fact that (5) and (6) are nonlinear implies that the linear analysis given will not hold far away from the equilibrium point and that in all likelihood the nonlinearity will result in this point being, in fact, a weakly stable or unstable focus rather than a center. Thus the long-term solutions cannot be predicted from linearizing the equations. However, it is the initial behavior of the model which is biologically relevant and, for a qualitative description of this, the elementary analysis adequately predicts the form of the solution.

As an example of the implications of this state of affairs we can consider $g(c') = A \exp(ac')$ and let $c'_0 = a = 1$ and let $a = [\text{conc}]^{-1}$ so that $A = 0.00015$; u, which is also arbitrary, can be set equal to 1 and the other parameters maintain the values they had in the previous model. Solving the

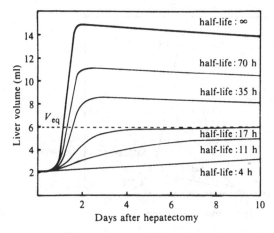

Figure 10.7. Theoretical growth for liver regeneration assuming that regeneration is controlled primarily by function and the consequent build up of an inhibitor. The thick line assumes that there is no secondary route for the degradation of the inhibitor (half-life: ∞); the other curves show how the liver is predicted to grow for a range of half-lives when the waste product can be degraded by a second route (from Bard, 1978).

equations (Figure 10.7: the case where the half-life of the stimulator = ∞, or its decay constant $q = 0$) shows that, within a few days of a 65% hepatectomy, the liver volume rises to about 14 ml and then slowly drops as cell death occurs. In the long term, it turns out that after about three months the liver has dropped below its equilibrium size and then starts to grow again; with other dose–response functions, the detailed solution would be different but such oscillations will always occur.

As an overshoot of liver size after regeneration has never been observed, it is clear that the simple model is wrong. It can be improved by suggesting that there is a second mechanism for degrading c' by, for example, giving it a decay constant q. Equation (6) becomes

$$\frac{dc'}{dt} = \frac{u(v_0 - v)}{v + w} - qc'. \tag{7}$$

This equation has a stable equilibrium point when $c' = 0$ but the dose–response function is so insensitive at the low values of c' which occur when $v \sim v_0$ that one can use the previous parameters to see what values of q are required to give a reasonable prediction for the system. It turns out (Figure 10.7) that when the decay constant is about 0.08 (half-life: 17 h) the liver returns to its original value in about eight days, a prediction not incompatible with the data. However, an examination of the mitotic indices implied by the model (Figure 10.8) shows that the long half-life demands a mitotic

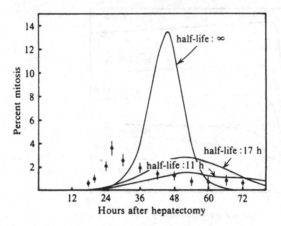

Figure 10.8. The mitotic indices predicted by the mitotic-stimulator theory assuming a delay of 15 h for the cell cycle (from Bard, 1978). Three half-lives for nonhepatic degradation of the waste product are shown: ∞ (no degradation); 17 h (optimal fit for growth;Figure 10.7); and 11 h. It can be seen that the optimal fit for growth does not match the data of Grisham (1962).

peak that not only occurs some 48 h after surgery (rather than the 26 h found in practice) but that the peak is far higher than that measured experimentally. The reason for this unexpectedly large amount of mitosis is that while the liver is small a backlog of waste product builds up.

The simple stimulator model is clearly unable to explain the experimental data. The addition of a second degradative route for the waste product improves the model but, at least in the form given here, leaves it still inadequate. The possibility that a still more sophisticated version would match the data cannot be excluded but it it probably correct to say that the model is unaceptable in the light of the current data.

Discussion
Analysis of the two models put forward to explain liver regeneration shows clearly that a liver-synthesized inhibitor can control the system while a waste product mitotic stimulator alone cannot. The possibility that a more complex regulatory system where for example a second organ assays some property of the liver and in turn regulates liver size, cannot be excluded, but there is no reason to invoke such a model until there is compelling evidence to exclude the simple inhibitor system. Indeed this latter system is compatible with much of the published data and in accordance with the intuitions of experimentors in the field (e.g. Sakai, 1970). One fact that the model does not explain, however, is the cellular hypertrophy

that occurs in the liver soon after hepatectomy and that seems to be independent of the later mitotic response (Tongendorff *et al.*, 1975). In the context of the waste product stimulator model, on the other hand, Bucher (1963) commented that there was no evidence to support the view that functional demand played a role in liver homeostasis and there has been no evidence since then to contradict her view.

There is some additional data that supports the inhibitor rather than the stimulator model of liver homeostasis. Fabrikant (1968) observed that, in the lobes remaining after hepatectomy, mitosis first occurred near arteries rather than veins. Assuming that a controlling molecule would be made in the liver and drain into such veins, the concentration would be higher there than near the arteries. Were the molecule a mitotic stimulator, one would then expect a higher mitotic rate near the vein; the fact that mitosis is higher in the periportal area around the artery is compatible with the molecule being an inhibitor.

An important aspect of any theory in biology is that it can make experimental predictions. One can use the inhibitor equations to predict the amount of mitosis expected if the dilution experiments of Glinos (1958) were repeated in more detail using stored plasma or saline to replace the plasma of a rat. The interested reader is referred to Bard (1978), where possible experiments to investigate the half-life and the dose–response curves of an inhibitor are considered.

There are two major problems with the analysis presented here: first, the considerable number of assumptions made may not all be valid. It is, for example, unlikely that the inhibitor concentration will be the same in the liver and in the blood or that all cells take exactly the same time to complete their cycle. These and other assumptions could be allowed for by expanding the equations and the complexity of the stimulation. No useful purpose would be served by this, for the additional number of unmeasurable and arbitrary parameters that would have to be introduced would give too much freedom to the system and allow the data to be matched with an inadequate number of constraints.

The second and perhaps more interesting difficulty lies in the nature of the inhibitor–liver interaction. This is concealed in the function $f(c)$ and it is assumed that it is biochemically easy to make an interaction described by a negative exponential. This is probably so (e.g. Walter, Parker & Yčas, 1967) but the mathematical approach used here gives no clue as to the details of the interaction. It is, however, worth repeating that the inhibitor is a member of the class of molecules known as 'chalones' which are tissue-specific mitotic inhibitors. Such molecules have been discussed at some length but there is in most cases a shortage of hard data on their existence in tissues where they would be expected to exist (see Forscher & Houck, 1973, for review) even though their presence has been discussed for over thirty years (see e.g. Weiss & Kavanau, 1957).

References

Alison, M. R. (1986). Regulation of hepatic growth. *Phys. Rev.* **66**, 499–541.

Bard, J. B. L. (1978). A quantitative model of liver regeneration in the rat. *J. Theoret. Biol.* **73**, 509–30.

Bard, J. B. L. (1979). A quantitative theory of liver regeneration in the rat. II. Matching an improved mitotic inhibitor model to the data. *J. Theoret. Biol.* **79**, 121–36.

Bucher, N. L. R. (1963). Regeneration of mammalian liver. *Int. Rev. Cytol.* **15**, 245–300.

Bullough, W. S. (1962). Control of mitotic activity in adult mammalian tissue. *Biol. Rev.* **37**, 307–42.

Fabrikant, J. I. (1968). The kinetics of cellular proliferation in regenerating liver. *J. Cell Biol.* **36**, 551–65.

Forscher, B. K. & Houck, J. C. (1973), eds. *Chalones: Concepts and Current Research*, Nat. Canc. Inst. Mon. No. 38.

Glinos, A. D. (1958). The mechanism of liver growth and regeneration. In *The Chemical Basis of Development*, ed. W. D. McElroy and B. Glass, Baltimore, Johns Hopkins Press, pp. 813–39.

Goss, R. J. (1964). *Adaptive Growth*, London, Logos Press.

Grisham, J. W. (1962). A morphological study of DNA synthesis and cell proliferation in regenerating rat liver. *Cancer Res.* **22**, 842–9.

LaBrecque, D. R. A. (1978). Diurnal rythm: effects on hepatic regeneration and hepatic regenerative stimulator substance. *Science* **199**, 1082–4.

McMahon, J. B., Farrelly, J. G. & Iype, P. T. (1982). Purification and properties of a rat liver protein that specifically inhibits the proliferation of nonmalignant epithelial cells from rat liver. *Proc. Nat. Acad. Sci.* **79**, 456–60.

Rabes, H. M., Wirshing, R., Tuczek, H. V. & Iseler, G. (1976). Analysis of cell cycle compartments of hepatocytes after partial hepatectomy. *Cell Tiss. Kinet.* **9**, 517–32.

Sakai, A. (1970). A humoral factor triggering DNA synthesis after partial hepatectomy in the rat. *Nature, Lond.* **228**, 1186–7.

Schulte-Hermann, R. (1977). 2 stage control of cell proliferation induced in rat liver by α-hexachlorocyclohexane. *Cancer Res.* **37**, 166–71.

Tongendorff, J., Trebin, R. & Ruhenstroth-Bauer, G. (1975). Critique of the 'critical mass' hypothesis of the regeneration of liver cells, *Amer. J. Pathol.* **80**, 519–24.

Verly, W. G., Deschamps, Y. & Pushpathadam, J. (1971). The hepatic chalone I: Assay method for the hormone and purification of the rabbit liver chalone. *Can. J. Biochem.* **49**, 1376–83.

Walter, C., Parker, R. & Yčas, M. (1967). A model for binary logic in biochemical systems. *J. Theoret. Biol.* **15**, 208–17.

Weiss, P. & Kavanau, J. L. (1957). A model of growth and growth control in mathematical terms. *J. Gen. Physiol.* **41**, 1–47.

11

Chaos

Simple models of biological systems have generally been expected to exhibit simple dynamic behavior. However, recently simple models have been developed which exhibit surprisingly complex behavior. For example, Mackey & Glass (1977) suggested the following model for hematopoiesis (blood cell differentiation). Let P be the concentration of mature circulating blood cells. Assuming that these cells are produced in the bone marrow and that there is a significant time delay, τ, between the initiation of cellular production in the bone marrow and the release of the mature cells in the blood, they write the following differential-delay equation to describe the changes in P:

$$\frac{dP(t)}{dt} = \left[\frac{\beta_0 \theta^n P(t-\tau)}{\theta^n + P(t-\tau)}\right] - \gamma P, \tag{1}$$

where β_0, θ, n and γ are constants. Numerical studies show that for small values of τ, P approaches a stable equilibrium value. However, if τ is increased, numerical solutions to (1) show that the blood cell population level oscillates, instead of remaining constant (see Figure 11.1a). Further increases in τ produce lower frequency oscillations with periods of 2, 4, 8 and 16 times the original period, as well as population changes that show no apparent periodicity, i.e. are **aperiodic** or **chaotic** (see Figure 11.1b). One can rigorously show that if no delay were present in (1), i.e. $\tau = 0$, or in any other first order ordinary differential equation, then the dynamics would be relatively simple. This conclusion is based on the fact that Smale (1967) proved that it is only for systems of three or more first order differential equations that chaotic behaviors appear. (In more precise technical language, Smale demonstrated that only systems with a dimension of at least three can have a new type of stable attractor, called a **strange attractor**.) Actually, it had been known since the work of Poincaré and Bendixson that systems of one or two first-order differential equations had only two types of attractors, fixed points and limit cycles.

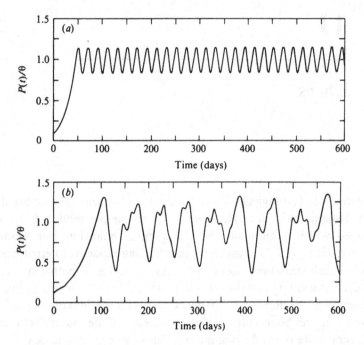

Figure 11.1. Numerical solutions to (1). (*a*) $P(0) = 0.1$, $\gamma = 0.1$ day^{-1}, $\beta_0 = 0.2$ day^{-1}, $n = 10$, and $\tau = 6$ days. The period of oscillation is 20 days. (*b*) Same as (*a*) except $\tau = 20$ days. Motion is now aperiodic. (From Mackey & Glass, 1977; copyright © 1977 by the American Association for the Advancement of Science.)

In order to understand how complicated dynamics can arise without involving ourselves in the difficult analysis of systems of three or more nonlinear differential equations, we shall examine an example from population biology developed in terms of a first-order difference equation. Although this is an aside from the main mathematical topic of ordinary differential equations in this volume, difference equations are used extensively in population biology, genetics and epidemiology (Hoppensteadt, 1975, 1976; May & Oster, 1976) as well as in numerical analysis, and hence they are of interest in their own right.

Consider a seasonally breeding population the generations of which do not overlap. Many natural populations, particularly temperate zone insects, are of this type. The way in which the population changes from generation to generation may be expressed in the general form

$$x_{t+1} = f(x_t), \qquad t = 0, 1, 2, \ldots, \tag{2}$$

where x_t is related to the size of the population in generation t. The function $f(x)$ will usually be nonlinear, that is, exhibit what an ecologist calls **density dependence**. Equation (2) is then a first-order, nonlinear difference equation.

A number of such models, employed in the ecological literature, have the property that x tends to increase from one generation to the next when it is small, but decreases when it is large. Thus $f(x)$ contains a single maximum. Also it is typical for the terms in $f(x)$ to have a common factor x, so one can express $f(x)$ as $f(x) = xg(x)$. This implies $f(0) = 0$, i.e. if a population vanishes one year it remains zero forever after. A specific example is the 'logistic' difference equation

$$N_{t+1} = rN_t[1 - (N_t/\theta)],$$

which can be written in the form

$$x_{t+1} = f(x_t) = rx_t(1 - x_t) \tag{3}$$

by the substitution $x = N/\theta$. It is this simple equation which we shall study.

For a fixed value of the parameter r, one can plot x_{t+1} versus x_t and obtain a graph of the function f as shown in Figure 11.2. Since only nonnegative populations are of interest, we restrict the values of x to the interval $0 \leqslant x \leqslant 1$. On this interval f is symmetrical about $x = \frac{1}{2}$, i.e.

$$f(\tfrac{1}{2} + y) = r(\tfrac{1}{2} + y)(\tfrac{1}{2} - y) = f(\tfrac{1}{2} - y), \qquad 0 \leqslant y \leqslant \tfrac{1}{2}.$$

Moreover f has a maximum value of $r/4$ at $x = \frac{1}{2}$, as we see overleaf:

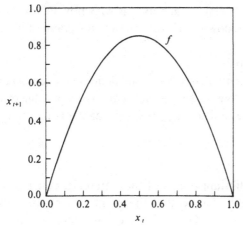

Figure 11.2. The graph of the function f defined by (3), for $r = 3.4$.

$$df/dx = r(1 - x) - rx = r(1 - 2x) = 0, \tag{4}$$

implying f has zero slope at $x = \frac{1}{2}$. Furthermore,

$$\left. \frac{d^2f}{dx^2} \right|_{x=\frac{1}{2}} = -2r < 0, \tag{5}$$

so $x = \frac{1}{2}$ is a maximum. From (3), $f(\frac{1}{2}) = r/4$. Since $x_{t+1} = f(x_t)$, the maximum value of $f(x)$ must be less than or equal to 1. Therefore we restrict our attention to $0 \leqslant r \leqslant 4$. For $0 \leqslant x \leqslant 1$, $x(1 - x) < 1$ and consequently if $r \leqslant 1$, x continually decreases, asymptotically approaching $x = 0$. For nontrivial dynamic behavior we require $1 < r \leqslant 4$.

Dynamics

The dynamics of a population can easily be computed from the graph of f as shown in Figure 11.3a. For a given x_0 draw a line upward until it intersects the graph of f. At this point draw a horizontal line to the x_{t+1}-axis and read off the value of x_1. To obtain the next value of x we repeat this process beginning with x_1 on the x_t-axis. A simple graphical method of locating the position of x_1 on the x_t-axis is to draw the 45° line, $x_t = x_{t+1}$. Then follow the horizontal line from x_1 to the 45° line. At the intersection point draw a vertical line downward and mark the value of x_1 on the x_t-axis. To obtain x_2 draw a vertical line upward from x_1 until it intersects the graph of f. Then draw a horizontal line to the x_{t+1}-axis and read off the value of x_2. Continuing in this way the sequence $x_0, x_1, x_2, x_3, \ldots$ can be obtained. This graphical procedure can be simplified by observing that every path from the 45° line to an axis is traversed twice in opposing directions. Eliminating these steps, as shown in Figure 11.3b, yields a rapid method of following trajectories. Further by adding a time axis as shown in Figure 11.3c and projecting the values of x_t downward one can obtain a graph of x_t versus t.

Equilibrium points

Just as for differential equations, one defines an **equilibrium point** (or **fixed point** or **steady state**) of a difference equation as a value of x which remains constant in time. Setting $x_{t+1} = x_t = x^*$ and solving the resulting algebraic equation

$$x^* = f(x^*) \tag{6}$$

yields the possible equilibrium points of the difference equation (2). An equivalent graphical method is to find the points where the curve $f(x)$ intersects the 45° line, $x_{t+1} = x_t$. This is illustrated in Figure 11.4. For (3) one must solve

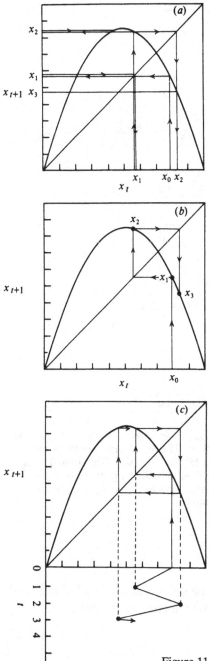

Figure 11.3. (*a*) and (*b*) graphical methods of determining x_t, $t = 1, 2, 3, \ldots$, as described in the text. (*c*) A graphical method of generating a plot of x_t versus t.

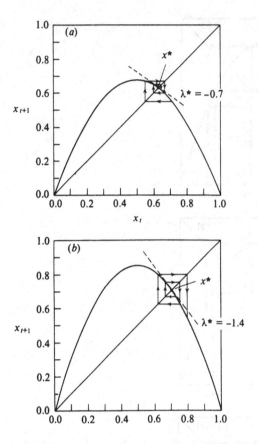

Figure 11.4. The graph of a function f defined by (3) with $r = 2.7$ (a) and $r = 3.4$ (b) (adapted from May, 1976). The fixed points are the places where the curve $f(x)$ intersects the 45° line, $x_{t+1} = x_t$. The dashed lines indicate the slope λ^* of $f(x)$ at the fixed points. In (a) the slope λ^* is between -1 and $+1$ and the fixed point is stable; for (b) the slope λ^* is less than -1, and the point is unstable.

$$x^* = rx^*(1 - x^*).$$

As is easily seen there are two equilibrium points: the trivial solution $x^* = 0$, which we shall not deal with further, and the nontrivial solution

$$x^* = 1 - (1/r). \qquad (7)$$

Stability

If a trajectory that starts near the equilibrium point x^* converges to x^*, then x^* is said to be **locally stable**. For a one-dimensional system, the stability of an equilibrium point depends on the slope of f at x^*. We denote this slope by λ^*, i.e.

$$\lambda^* \equiv \left.\frac{df}{dx}\right|_{x=x^*} \tag{8}$$

As Figure 11.4 illustrates, if the slope of f at x^* lies between $45°$ and $-45°$ (that is if $-1 < \lambda^* < 1$), then the equilibrium point will be locally stable, attracting all trajectories in its neighborhood. For (3), we see using (4) and (7) that

$$\lambda^* = r(1 - 2x)|_{x=x^*=1-(1/r)} = 2 - r. \tag{9}$$

Thus the equilibrium point x^* is stable if and only if $1 < r < 3$.

To derive this stability result in an analytical fashion, write

$$x_t = x^* + x_t',$$

where x_t' denotes a small excursion from x^*. Then using a Taylor series approximation for $f(x^* + x_t')$, (3) becomes

$$x^* + x_{t+1}' = f(x^*_\bullet + x_t') \approx f(x^*) + \lambda^* x_t'. \tag{10}$$

Since $x^* = f(x^*)$, (10) becomes

$$x_{t+1}' = \lambda^* x_t'. \tag{11}$$

Consequently, if $|\lambda| > 1$, $|x_t'| \to \infty$ as $t \to \infty$, whereas, $|x_t'| \to 0$ as $t \to \infty$, if $|\lambda^*| < 1$. When $|x_t'| \to 0$, $x_t \to x^*$ and x^* is said to be locally stable.

What happens when $r > 3$? To answer this question, it is useful to examine the map which relates populations two generations apart, i.e. x_{t+2} and x_t. By iterating (2), one sees

$$x_{t+2} = f[f(x_t)]; \tag{12}$$

or introducing a simplifying notation

$$x_{t+2} = f^{(2)}(x_t). \tag{13}$$

For the function defined by (3),

$$x_{t+2} = rx_{t+1}(1 - x_{t+1}) = r^2 x_t(1 - x_t)[1 - rx_t(1 - x_t)]. \tag{14}$$

Fixing the value of r, one can plot x_{t+2} versus x_t and obtain a graph of $f^{(2)}$. Notice that just as f was symmetric about $\frac{1}{2}$, $f^{(2)}$ must be symmetric about $\frac{1}{2}$ since x_t only appears in (14) in the combination $x_t(1 - x_t)$. The maxima and minima of $f^{(2)}$ can be obtained by differentiation. Using the chain rule

$$\frac{df^{(2)}}{dx} = \frac{df}{dx}\bigg|_{x=f(x)} \frac{df}{dx}. \tag{15}$$

From (4),

$$df^{(2)}/dx = r[1 - 2f(x)]r(1 - 2x) = r^2[1 - 2rx(1 - x)][1 - 2x]. \tag{16}$$

A point p such that

$$\frac{df}{dx}\bigg|_{x=p} = 0$$

is called a *critical point* of f. From (16) we see that the critical points of $f^{(2)}$ are $x = \frac{1}{2}$ and any real solutions to the equation

$$1 - 2rx + 2rx^2 = 0, \tag{17}$$

i.e.

$$x = \frac{2r \pm \sqrt{[4r(r-2)]}}{4r}. \tag{18}$$

Thus for $r < 2$, $x = \frac{1}{2}$ is the only critical point. For $r > 2$ there are three critical points: $x = \frac{1}{2}$ and the two values of x given by (18). One can further show that for $r < 2$, $x = \frac{1}{2}$ is a maximum, whereas for $r > 2$, $x = \frac{1}{2}$ is a minimum and the two values of x given by (18) are maxima. In Figure 11.5, $f^{(2)}$ is plotted for various values of r.

Fixed points of period 2

Fixed points of period 2 are points invariant under two iterations of the map f. These points, written $(x^*)^{(2)}$, can be found either algebraically by solving the equation

$$(x^*)^{(2)} = f^{(2)}[(x^*)^{(2)}], \tag{19}$$

or graphically from the intersection of the map $f^{(2)}(x)$ with the 45° line as shown in Figure 11.5. Clearly, the equilibrium point x^* of (6) is a solution of (19). A fixed point of period 1 is a degenerate case of a fixed point of period 2.

The stability of a fixed point of period 2 is determined by the slope of $f^{(2)}(x)$ at the point $x^{*(2)}$. By the chain rule

$$\frac{d}{dx} f[f(x)]\bigg|_{x=(x^*)^{(2)}} = \frac{df}{dx}\bigg|_{x=f[(x^*)^{(2)}]} \frac{df}{dx}\bigg|_{x=(x^*)^{(2)}} \tag{20}$$

For the degenerate fixed point $(x^*)^{(2)} = x^*$, $f(x^*) = x^*$ and thus

$$\frac{df^{(2)}}{dx}\bigg|_{x=x^*} = \left[\frac{df}{dx}\bigg|_{x=x^*}\right]^2 = (\lambda^*)^2. \tag{21}$$

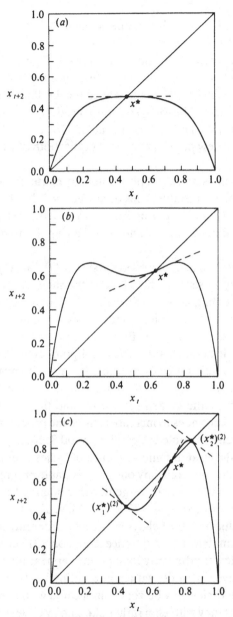

Figure 11.5. The relationship between x_{t+2} and x_t as described by (14) for various values of r (from May, 1976). (a) $r = 1.9$. For $r \leqslant 2$, $f^{(2)}$ has a single hump and only one fixed point. (b) $r = 2.7$. When $r > 2$, $f^{(2)}$ has two humps. The basic fixed point x^* is stable if $1 < r < 3$. (c) $r = 3.4$. The 45° line intersects $f^{(2)}$ in three places when $r > 3$. The basic fixed point x^* is unstable, since the slope of $f^{(2)}$ at x^* is greater than 1. The two new fixed points, $(x_1^*)^{(2)}$ and $(x_2^*)^{(2)}$, are stable for this value of r giving rise to cycles of period 2.

This fact can now be used to see what happens when the fixed point x^* becomes unstable. If $|\lambda^*| < 1$ (i.e. $r < 3$) then x^* is stable, and from (21), the slope of $f^{(2)}(x)$ at x^* must lie between $0°$ and $45°$ as shown in Figure 11.5. For slopes in this range, $f^{(2)}$ will only intersect the $45°$ line once as shown in the figure. At $|\lambda^*| = 1$, i.e. $r = 3$ in (3), $f^{(2)}$ is tangent to the $45°$ line at x^*. For $|\lambda^*| > 1$ ($r > 3$), the humps of $f^{(2)}$ become so pronounced that the slope of $f^{(2)}$ at x^* is steeper than $45°$. Thus x^* is unstable. However, $f^{(2)}$ now intersects the $45°$ line at two new points $(x_1^*)^{(2)}$ and $(x_2^*)^{(2)}$ both of which are stable.

In summary, as $f(x)$ becomes more steeply humped because of increases in r, the fixed point x^* becomes unstable. This occurs at $r = 3$. For $r > 3$ two new and initially stable fixed points of period 2 arise. In time the system then alternates between the fixed points $(x_1^*)^{(2)}$ and $(x_2^*)^{(2)}$ giving rise to periodic behavior.

The stability of the period 2 cycle depends on the slope of $f^{(2)}$ at $(x_1^*)^{(2)}$ and $(x_2^*)^{(2)}$. From (20) it is easy to see that the slope is the same at these two points. This stability-determining slope has the value $\lambda = 1$ at the birth of the 2-point cycle and then decreases through zero towards $\lambda = -1$ as the hump in $f(x)$ steepens with increasing values of r. Beyond the point $\lambda = -1$ the 2-point cycle becomes unstable and generates a stable 4-point cycle. Examining $f^{(4)}$ one can show that in this range of r values, $f^{(4)}$ has 4 humps which steepen as r increases. The slope of $f^{(4)}$ at the intersection with the $45°$ line changes from $+45°$ to $-45°$ as the kinks in $f^{(4)}$ grow, and thus the 4-point cycle goes unstable. As one continues to increase r, the same process repeats itself giving rise to a hierarchy of stable cycles with period 8, 16, 32, ..., 2^n. Although this process produces an infinite sequence of cycles with periods 2^n, $n \to \infty$, the range of r values wherein any one cycle is stable progressively decreases, so that the entire process is convergent, being bounded above by some critical value, r_c, of r. For (3), $r_c = 3.5700 \ldots$ (May, 1976).

Beyond the limiting value r_c, the behavior is quite different, and we suggest that the reader simulate the difference equation (3) on a hand calculator to see this. Cycles of arbitrarily long periods appear after r_c. At first these cycles all have even periods, with x_t alternating between values above and below x^*. These cycles may be very complicated with periods of several thousand points, but they will seem rather like a noisy cycle of period 2. As r continues to increase, the first odd period cycle appears at $r = 3.6786$ ... (May, 1976). Initially these cycles have very long odd periods, but as r increases, cycles with smaller and smaller odd periods appear, until a 3-point cycle appears at $r = 3.8284 \ldots$. Beyond this point, cycles of all integer periods are present, i.e. changing the initial point x_0 gives rise to cycles of all lengths. This is known as Sarkovskii's Theorem (cf. Devaney, 1989). But even more surprisingly, there are an uncountable number of initial points x_0 which give totally *aperiodic* trajectories; no matter how many points x_t are generated, the pattern never repeats (see Figure 11.6). This situation where an infinite number of different orbits can occur has been christened '*chaotic*' by Li & Yorke (1975).

This is not meant to be a rigorous definition of chaos. In fact, there are many different definitions of chaos, ranging from measure theoretic notions of randomness in ergodic theory, to topological definitions (Devaney, 1989). Generally, a mapping is said to be chaotic if periodic points are dense, there are points which eventually move under iteration of the map from a small neighborhood to any other neighborhood, and there is a sensitive dependence on initial conditions (Devaney, 1989).

In the chaotic region a single trajectory has a completely well-defined and deterministic behavior. However, slight changes in the initial conditions can give rise to very different long-term behavior. If one observed a process governed by (3) with r in the chaotic regime the orbit might look indistinguishable from one generated by a stochastic process. Numerical simulations tend to confirm this (May, 1976). Other difference equations with nonanalytic functions $f(x)$, such as

$$x_{t+1} = \begin{cases} rx_t & \text{if } x_t < \frac{1}{2} \\ r(1-x_t) & \text{if } x_t > \frac{1}{2}, \end{cases} \tag{22}$$

can be proven to have truly random behavior (cf. May, 1976). For (3) one

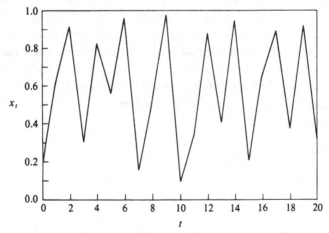

Figure 11.6. The solution to (3) with $r = 3.9$ and $x_0 = 0.2$ for $t = 0$, $1, 2, \ldots, 20$. No value occurs twice and the pattern does not repeat itself.

can show that for any specified parameter value there is one unique cycle that is stable, and attracts essentially all initial points (Smale & Williams, 1976). The remaining infinite number of other cycles and uncountable number of aperiodic trajectories, occur for a set of initial conditions that have measure zero. However, since any particular stable cycle is likely to occur only for an extraordinarily narrow range of r values, and because of the long time one expects to elapse before the transients associated with any given initial condition damp out, in practice the unique cycle is unlikely to be detected by observation. Thus a stochastic description of the dynamics is likely to be appropriate even though the underlying process is deterministic. This has been well appreciated in statistical mechanics for very large systems. What is new here is that *very simple* deterministic dynamics can give rise to a situation that looks stochastic. The implications of this for ecology and other areas of science are most unsettling. From a modeling point of view, it means that it may be impossible to distinguish data generated by a simple deterministic process, such as (3), from stochastic noise or experimental error in measurement. Alternatively, phenomena that look chaotic may come from simple underlying dynamic laws. Ruelle & Takens (1971) have taken this viewpoint in trying to explain turbulence in fluids.

Other examples and applications

In the chaotic regime arbitrarily close initial conditions can lead to trajectories, which after a sufficiently long time, diverge widely. Thus even with a simple model in which all parameters are determined exactly, long-term prediction is impossible. Lorenz (1963, 1964), in attempting to predict weather, noticed this phenomenon and called it the '*butterfly effect*': assuming one could describe the atmosphere exactly by a deterministic model, the fluttering of a butterfly's wings could change the initial conditions and in the chaotic regime alter the long-term predictions of the model. Besides noticing the chaotic behavior of (3), Lorenz (1963) also discovered a system of three ordinary differential equations which apparently also exhibit chaos. These equations are rather simple:

$$dx/dt = -ax + ay, \tag{23}$$

$$dy/dt = xz + bx - y, \tag{24}$$

$$dz/dt = xy - cz. \tag{25}$$

For $a = 10$, $b = 28$ and $c = \frac{1}{3}$ aperiodic behavior is observed in numerical simulations.

A simple mechanical system composed of a double pendulum with nonlinear springs, gives rise to chaotic motion. A picture of the orbits is given by Arnold & Avez (1970).

Turbulence in fluid motion has been proposed to occur via a sequence of bifurcations, similar to those we described for difference equations (see Ruelle & Takens, 1971).

Chemical reactions have recently been discovered that seem to exhibit chaotic behavior. Olsen & Degn (1977) describe this effect in the oxidation of NADH by O_2 catalyzed by horseradish peroxidase. Schmitz, Graziani & Hudson (1977) observed what appears to be chaotic states in the Belousov–Zhabotinsky reaction run in an isothermal continuous flow stirred tank reactor. Schmitz has also observed chaotic behavior in an experimental study of hydrogen oxidation on a platinum catalyst. Rössler (1977) has suggested abstract kinetic systems which exhibit chaos.

A method of discovering chaotic behavior

The existence of aperiodic behavior may be established for any first-order difference equation by utilizing a theorem of Li & Yorke (1975). Starting at the critical point, b, in Figure 11.7, trace out the next two iterates of the map, $c = f(b)$ and $d = f(c)$, as well as the pre-image a, $b = f(a)$. If $a \geq d$ (i.e. if d is to the left of a) then there exists an infinite number of periodic points and the function can generate aperiodic motions.

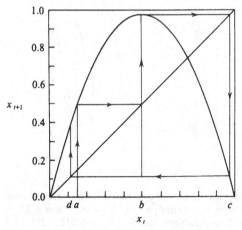

Figure 11.7. Method of showing the existence of chaotic solutions. If $d \leq a$ then a period-3 cycle exists and there are initial conditions which generate aperiodic trajectories.

This procedure was used experimentally by Olsen & Degn (1977). From experimental observations of the O_2 concentration in a peroxidase-catalyzed oxidation reaction of NADH (Figure 11.8), they extracted the amplitudes of the oscillations. Plotting each amplitude, they obtained a transition map f, and utilized the Li & Yorke procedure to show that chaotic behavior existed. This is illustrated in Figure 11.9.

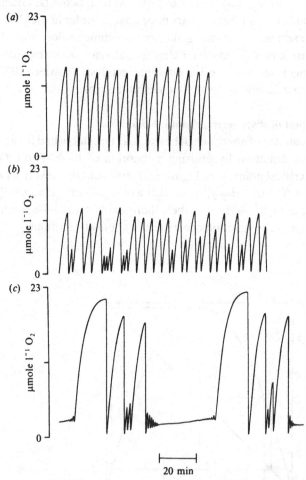

Figure 11.8. Oscillations in O_2 concentration in the peroxidase-catalyzed oxidation of NADH in a system open to O_2 (from Olsen & Degn, 1977). The overall peroxidase-catalyzed reaction is $2NADH + 2H^+ + O_2 \rightarrow 2NAD^+ + 2H_2O$. The concentration of peroxidase was varied in the three experiments shown; its value being (a) $0.90\,\mu$mole l^{-1}; (b) $0.55\,\mu$mole l^{-1} and (c) $0.45\,\mu$mole l^{-1}.

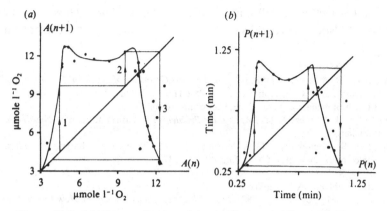

Figure 11.9. (a) The amplitudes of the oscillations in Figure 11.8b $A(n + 1)$, plotted against the preceding amplitudes $A(n)$. (b) The oscillation periods in Figure 11.8b. $P(n + 1)$ plotted against the preceding periods, $P(n)$. Trajectories with arrows were drawn and show that the transition functions allow a period-3 cycle. (From Olsen & Degn, 1977.)

Conclusion

For mathematical simplicity, the bulk of our discussion in this section has concerned a difference equation of the type found in population dynamics. We have, however, given specific examples of apparently chaotic behavior in chemical systems. Moreover, the system of (23)–(25) is of the type that governs interacting cell populations, so the chaotic solutions of these equations might be expected to find counterparts in cellular biology. In general, suitable combinations of nonlinearity, delay, and high-order systems promote chaotic behavior, so one expects more and more instances of such behavior to reveal themselves in biology.

Since this chapter was first written in 1979 there has been an explosion of research activity in nonlinear dynamics and the study of chaotic systems. A number of excellent textbooks are now available (cf. Abraham & Shaw, 1982; Guckenheimer & Holmes, 1983; Bergé, Pomeau & Vidal, 1984; Thompson & Stewart (1986), Shuster, 1988; Wiggins, 1988; Devaney, 1989), as well as more popular accounts (Gleick, 1987; Peterson, 1988). Applications to various areas in biology have also been developed (see Holden, 1986 for reviews). For example, Schaffer (1985) and Schaffer & Knot (1986) examine applications in ecology and in the epidemiology of childhood diseases. Rapp (1986), Winfree (1987), Glass & Mackey (1988), Glass *et al.* (1986, 1988), Goldberger *et al.* (1986), and Kaplan *et al.* (1988) among others, apply these ideas to the dynamics of the heart and other physiological systems. Reports of chaos in immunological experiments

(Lundkvist *et al.*, 1989) and theories (Perelson, 1989; Stewart & Varela, 1989; De Boer, Kevrekidis & Perelson, 1990) are beginning to appear.

References

Abraham, R. & Shaw, C. (1982). *Dynamics: The Geometry of Behavior*. Part One: Periodic behavior. Part Two: Chaotic behavior. Part Three: Global behavior. Part Four: Bifurcation behavior, Santa Cruz, California, Aerial Press.

Arnold, V. I. & Avez, A. (1970). *Ergodic Problems in Classical Mechanics*, New York, Benjamin.

De Boer, R., Kevrekidis, I. G. & Perelson, A. S. (1990). A simple idiotype network model with complex dynamics. *Chem. Eng. Sci.* **45**, 2375–82.

Devaney, R. L. (1989). *An Introduction to Chaotic Dynamical Systems*, 2nd edn, Redwood City, California, Addison-Wesley.

Glass, L. & Mackey, M. C. (1988). *From Clocks to Chaos*, Princeton, New Jersey, Princeton University Press.

Glass, L., Shrier, A. & Bélair, J. (1986). Chaotic cardiac rhythms. In *Chaos*, Holden, A. V., ed., pp. 237–56, Princeton, New Jersey, Princeton University Press.

Glass, L., Beuter, A. & Larocque, D. (1988). Time delays, oscillations, and chaos in physiological control systems. In *Nonlinearity in Biology and Medicine*, Perelson, A. S., Goldstein, B., Dembo, M. & Jacquez, J. A., eds, pp. 111–25, New York, Elsevier.

Gleick, J. (1987). *Chaos: Making a New Science*, New York, Viking.

Goldberger, A. L., Bhargava,V., West, B. J. & Mandell, A. J. (1986). Some observations on the question: Is ventricular fibrillation 'chaos'? *Physica* **19D**, 282–9.

Guckenheimer, J. & Holmes, P. (1983). *Nonliner Oscillations, Dynamical Systems, and Bifurcations of Vector Fields*, New York, Springer-Verlag.

Holden, A. V. (1986). *Chaos*, Princeton, New Jersey, Princeton University Press.

Hoppensteadt, F. C. (1975). *Mathematical Theories of Populations: Genetics and Epidemics*, Philadelphia, SIAM.

Hoppensteadt, F. C. (1976). *Mathematical Methods of Population Biology*, New York, Courant Institute of Mathematical Sciences.

Kaplan, D. T., Smith, J. M., Saxberg, B. E. H. & Cohen, R. J. (1988). Nonlinear dynamics in cardiac conduction. In *Nonlinearity in Biology and Medicine*, Perelson, A. S., Goldstein, B., Dembo, M. & Jacquez, J. A., eds, pp. 19–48, New York, Elsevier.

Li, T.-Y. & Yorke, J. A. (1975). Period three implies chaos. *Amer. Math. Monthly* **82**, 985–92.

Lorenz, E. N. (1963). Deterministic nonperiodic flows. *J. Atmosph. Sci.* **20**, 130–41.

Lorenz, E. N. (1964). The problem of deducing the climate from the governing equations. *Tellus* **16**, 1–11.

Lundkvist, I., Coutinho, A., Varela, F. & Holmberg, D. (1989). Evidence for a functional idiotypic network among natural antibodies in normal mice. *Proc. natl. Acad. Sci. USA* **86**, 5074–8.

Mackey, M. C. & Glass, L. (1977). Oscillation and chaos in physiological control systems. *Science* **197**, 287–9.

May, R. M. (1976). Some mathematical models with very complicated dynamics. *Nature, Lond.* **261**, 459–67.

May, R. M. & Oster, G. (1976). Bifurcations and dynamic complexity in simple ecological models. *Amer. Natural.* **110**, 573–99.

Olsen, L. F. & Degn, H. (1977). Chaos in an enzyme reaction. *Nature, Lond.* **267**, 177–8.

Perelson, A. S. (1989). Immune network theory. *Immunol. Rev.* **110**, 5–36.

Peterson, I. (1988). *The Mathematical Tourist*. New York, Freeman.

Rapp, P. E. (1986). Oscillations and chaos in cellular metabolism and physiological systems. In *Chaos*, Holden, A. V., ed., pp. 179–208, Princeton, New Jersey, Princeton University Press.

Rössler, O. E. (1977). Chaos in abstract kinetics: two prototypes. *Bull. Math. Biol.* **39**, 275–89.

Ruelle, D. & Takens, F. (1971). On the nature of turbulence. *Commun. Math. Phys.* **20**, 167–92.

Schaffer, W. M. (1985). Can nonlinear dynamics elucidate mechanisms in ecology and epidemiology? *IMA J. Math. Appl. Med. Biol.* **2**, 221–52.

Schaffer,W. M. & Kot, M. (1986). Differential systems in ecology and epidemiology. In *Chaos*, Holden, A. V., ed., pp. 158–78, Princeton, New Jersey, Princeton University Press.

Schmitz, R. A., Graziani, K. R. & Hudson, J. L. (1977). Experimental evidence of chaotic states in the Belousov–Zhabotinskii reaction. *J. Chem. Phys.* **67**, 3040–4.

Shuster, H. G. (1988). *Deterministic Chaos*, 2nd revised edn, Weinheim, Fed. Republic of Germany, VCH Verlagsgesellschaft.

Smale, S. (1967). Differentiable dynamical systems. *Bull. Amer. Math. Soc.* **73**, 747–817.

Smale, S. & Williams, R. F. (1976). The qualitative analysis of a difference equation of population growth. *J. Math. Biol.* **3**, 1–4.

Stewart, J. & Varela, F. (1989). Exploring the meaning of connectivity in the immune network. *Immunol. Rev.* **110**, 37–61.

Thompson, J. M. T. & Stewart, H. B. (1986). *Nonlinear Dynamics and Chaos: Geometrical Methods for Engineers and Scientists*, New York, Wiley.

Wiggins, S. (1988). *Global Bifurcations and Chaos: Analytical Methods*, New York, Springer-Verlag.

Winfree, A. T. (1987). *When Time Breaks Down: The Three-Dimensional Dynamics of Electrochemical Waves and Cardiac Arrhymias*, Princeton, New Jersey, Princeton University Press.

INDEX

Printed in the United States
By Bookmasters